Presenting and Representing Environments

The GeoJournal Library

Volume 81

The titles published in this series are listed at the end of this volume.

Presenting and Representing Environments

Edited by

GRAHAM HUMPHRYS

Former Senior Lecturer in Geography,
University of Wales Swansea, UK

and

MICHAEL WILLIAMS

Emeritus Professor of Education,
University of Wales Swansea, UK

Springer

A C.I.P. Catalogue record for this book is available from the Library of Congress

ISBN-10 1-4020-3813-5 (HB)
ISBN-13 978-1-4020-3813-6 (HB)
ISBN-10 1-4020-3814-3 (e-book)
ISBN-13 978-1-4020-3814-3 (e-book)

Published by Springer,
P.O. Box 17, 3300 AA Dordrecht, The Netherlands.

www.springeronline.com

Printed on acid-free paper

Printed in the Netherlands.

CONTENTS

Acknowledgements vii
List of contributors ix

1. Cross-disciplines, Cross-cultures: the Environment as Social Construction 1
Graham Humphrys and Michael Williams

2. Environmentalism *qua* Environmental Non-Government Organisations and the Contested Remapping of British Columbia's Forests 17
Roger Hayter

3. Re-Negotiating Science in Protected Areas: Grizzly Bear Conservation in the Southwest Yukon 33
Douglas Clark and D. Scott Slocombe

4. The Moorlands of England and Wales: Histories and Narratives 55
Ian G. Simmons

5. Exploration Literature and the Canadian Environment: From Way-Finding to Ways of Representation and Reading 69
Wayne K. D. Davies

6. Changing Public Participation and the Environment of Swansea East 89
Graham Humphrys

7. Sustaining Local Riverine Environments: the River Valleys Committee in Calgary, Alberta, Canada 105
Dianne Draper

8. A Picnic in March: Media Coverage of Climate Change and Public Opinion in The United Kingdom 121
Tammy Speers

9. Challenging the Negative Critique of Landscape 137
Robert A. Newell

10. Threatened Environments, Atrophying Cultures, Lacklustre Policies 153
Colin H. Williams

11. Sustaining Arctic Visions, Values and Ecosystems: Writing Inuit Identity, Reading Inuit Art in Cape Dorset, Nunavut 167
Nancy C. Doubleday

12. Cultivating a New Cattle Culture: Lifelong Learning and Pasture Land Management 181
Ian Maclachlan, Nancy G. Bateman and Thomas R. R. Johnston

13. Environmental Education and Lifelong Learning: Awareness to Action 197
Michael Williams

Index 215

ACKNOWLEDGEMENTS

For Chapter 2, the author is grateful for the financial support of SSHRC, specifically grant # 410 2001 0071, in carrying out the research which provided input for this chapter.

The research reported in Chapter 3, undertaken in Kluane, was financially supported by the Social Sciences and Humanities Research Council of Canada, Wilfrid Laurier University, a Canon National Park Science Scholarship, Mountain Equipment Co-op, the Department of Indian Affairs and Northern Development's Northern Scientific Training Program, a TransCanada Pipelines Graduate Award, and Yukon College's Northern Research Institute. The authors gratefully acknowledge the contributions of their interview participants, and the efforts of Linaya Workman, Lyn Hartley, and many other individuals who have facilitated their fieldwork. The input of several anonymous reviewers substantially helped the development of the manuscript. The views expressed are the authors' own, and in no way should be considered to represent those of any organisation mentioned in the chapter.

For Chapter 6, the author gratefully acknowledges the valuable information provided and the helpful comments on earlier drafts of the paper by Donald Anderson, Member of Parliament for Swansea East since 1974, and Dorothy Anderson, PhD, BSc.

For Chapter 7, the author would like to thank Robin Poitras for preparing the Figures, and Bill Morrison for his insightful comments on an earlier draft of this chapter.

For Chapter 12, the authors wish to acknowledge the assistance of Bob Dobson and Lorne Fitch (Provincial Riparian Specialist, Alberta Riparian Habitat Management Society) who graciously consented to review the manuscript at short notice and made helpful suggestions and corrections.

The editors acknowledge the generous financial support of the Canadian Studies in Wales Group, the Academic Office of the Canadian Federal Government and of the Gregynog Fund. Through these agencies it was possible to bring the authors of this book to a seminar held in the University of Wales Conference Centre at Gregynog in

March 2004. We also acknowledge the additional financial support they provided towards the cost of copy-editing and rendering the manuscript camera-ready for publication.

Finally, the editors wish to thank Kate Williams for preparing the manuscript for publication.

LIST OF CONTRIBUTORS

Nancy Bateman owns and operates NGB Evaluation Solutions Inc. in Lethbridge, Alberta. A human geographer, she specialises in the structured assessment of awareness programmes that promote environmental stewardship. Her focus includes riparian and water quality issues relating to agriculture, rangeland management, recreational lake management, urban residential yard practices, and watershed planning.

Douglas Clark is a doctoral candidate in the Department of Geography and Environmental Studies, Wilfrid Laurier University, Waterloo, Ontario, Canada. His research interests broadly include institutions and governance for wildlife and ecosystems. He has 12 years of professional experience with Parks Canada in a variety of positions and locations.

Wayne Davies is a native of Pontypridd and was educated at the University College of Wales (Aberystwyth). Before becoming Professor in Calgary he held lecturing positions in the Universities of Southampton and Swansea. He is author-editor of 10 books and over 120 academic articles or chapters, mainly in urban geography, and is an Honorary Fellow of the University of Swansea.

Nancy C. Doubleday is Associate Professor of Geography and Environmental Studies at Carleton University. She works at the interface of environmental change and ecosystem approaches to decision-making, integrating physical, ecological, social and cultural understandings. She has collaborated with northern peoples since 1980.

Dianne L. Draper is a Professor in the Geography Department at the University of Calgary where she is an environmental and resource management specialist. She has been a volunteer on Calgary's River Valleys Committee for more than a decade and served as chair from 2001-2005. Her current research focuses on governance, sustainability and quality of life issues in communities affected by tourism growth and its impacts on water resources, parks and protected areas.

Roger Hayter is a Professor of Geography at Simon Fraser University with special interests in industrial location dynamics, industrial organization, and regional development. Much of his work has focused on British Columbia's forest economy and he has also research interests in Pacific Asia centred on Japanese production systems.

Graham Humphrys has degrees from the universities of Bristol, McGill and Swansea and is a former Senior Lecturer in Geography and former Dean of the Faculty of Social Sciences at Swansea University. He has special interests in economic geography, the environment and geographical education.

Tom Johnston earned his Ph.D. in Geography at the University of Waterloo and is currently Associate Professor of Geography and Coordinator of Canadian Studies at The University of Lethbridge. Current research interests include the institutional arrangements surrounding water management and the response by dryland farmers to recent droughts in Western Canada.

Ian MacLachlan is an Associate Professor of Geography at The University of Lethbridge in southern Alberta. He is an industrial geographer with research interests in the integration of the cattle and beef commodity chain, the restructuring of industrial meat packing in Canada, and the locational dynamics of nineteenth century livestock slaughter in the United Kingdom.

Robert A. Newell, MA, BA Hons. is a landscape painter and a member of the Royal Cambrian Academy. He is a Lecturer in Fine Art at Swansea Institute of Higher Education, having studied at Wimbledon School of Art and Goldsmiths College, University of London.

Ian Simmons is an Emeritus Professor of the University of Durham, UK, following his retirement from a Chair of Geography in 2001. He also taught at the University of Bristol, York University Ontario, the University of Alberta and the University of California at Berkeley. His PhD (University College London 1962) was supplemented by a DLitt (Dunelm 1998) and an Honorary DSc conferred by the University of Aberdeen in 2005. He holds the Victoria Medal of the Royal Geographical Society and is a Fellow of the British Academy, of the Society of Antiquaries of London and of Academia Europaea. His books on the environmental history of Great Britain (2001) and of the Moorlands of England and Wales (2003) will, he hopes, be accompanied by a 'global' volume in the next few years.

Scott Slocombe is Professor in the Department of Geography and Environmental Studies, Wilfrid Laurier University, Waterloo, Ontario, Canada. He studied at the Universities of Victoria, British Columbia and Waterloo, and undertakes research in environmental planning, management, and assessment, protected areas, and systems approaches.

Tammy Speers is a Lecturer at the Cardiff School of Journalism, Media and Cultural Studies. She is co-author (with Professors Ian Hargreaves and Justin Lewis) of *Towards a Better Map: Science, the Public and the Media.* Her research interests include media coverage of health and science issues and the relationships between sources, public relations, journalists and the public.

Colin H. Williams is Research Professor, School of Welsh, Cardiff University, Adjunct Research Professor, Department of Geography, The University of Western Ontario, and Honorary Professor in Celtic Studies at the University of Aberdeen. He has long been active in promoting Canadian Studies in the UK. His current research focuses on issues of governance in Europe and Canada.

Michael Williams is an Emeritus Professor and former Dean of the Faculty of Education and Health Studies and Head of the Department of Education at Swansea University. He has special interests in geographical and environmental education, lifelong learning and curriculum evaluation. His recent edited publications include (with Rod Gerber) *Geography, Culture and Education* (Kluwer, 2002) and (with Graham Humphrys) *Citizenship Education and Lifelong Learning: Power and Place* (Nova, 2003).

1. CROSS-DISCIPLINES, CROSS-CULTURES: THE ENVIRONMENT AS SOCIAL CONSTRUCTION

GRAHAM HUMPHRYS AND MICHAEL WILLIAMS

Global environmental change has become an accepted fact as the effects are becoming more obvious. At the local level, previously simple environmental matters such as the availability and use of fuel, the availability of water supplies, and even the disposal of household waste, have become matters of concern to people all over the world. In response, most countries have agreed to adhere to global environmental objectives agreed at international conferences. They are passing new laws and issuing regulations that require people and businesses to meet often ambitious environmental standards. Even more far reaching is the attention being paid to the environment in the media. Radio, television and newspapers continually report events and matters of environmental concern, while widespread availability of the Internet has dramatically increased the facility with which people can access environmental knowledge and debate. As the environment has loomed larger in everyday life, the way in which it is presented and represented has become much more important and it is these two aspects which are the focus of this book. While presentation and representation influence environmental decisions and actions by individuals, they are especially important when policies and actions that affect a large number of people are being decided by public and private organisations.

Particularly important in the following chapters are the value-laden processes that are always involved whenever environmental information is presented and the environment represented. There are two reasons why such subjectivity always intrudes. The first relates to the input side. Knowledge of any aspect of the environment is almost always incomplete. This is particularly true of the processes at work so that there is an element of uncertainty both about what is happening at present and of the likely impact of any action taken. In addition, whenever environmental matters are considered there will be disputed values so that different parties will have different views about what should be done. Where commercial and financial profit and loss are at stake, these values influence the views expressed so that it is necessary to identify who will gain and who will lose from any proposed environmental change. This is especially important where the stakes are high.

G. Humphrys and M. Williams (eds.), Presenting and Representing Environments, 1-15.

The second reason for the subjectivity derives from how environmental presentation and representation are received. Individuals perceive environments through various filters and a particular environment will mean substantially different things to different individuals, while the same environment may be perceived by the same person in different ways at different times according to the filter selected for viewing. Individuals also have different perceptions and attitudes to the environment at different scales, with geographical proximity usually an important influencing factor. Previously received knowledge of the environment will at least colour the values of recipients and their attitudes and reactions to new information. Duncan and Ley (1993) put this very succinctly, "The reader understands a text by situating it within two interpenetrating fields of reference – the extra-textual, the reader's experiences in the world, and the inter-textual, the context of other texts" (p.9). This new view of 'reality' came to prominence in the late twentieth century. The earlier views that had emerged from the Enlightenment suggested that there is an objective world of which the environment is a major part that is awaiting us to explore and discover. Today, it is recognised that what we think of as 'reality' is a social construct that we create through our relationships and which is subject to modification and development over time by further inputs from outside ourselves interacting with our own thought processes (Hartley 1988). How the environment is presented and represented will certainly influence perceptions and attitudes, and through these decisions and actions, though the nature of the influence will depend upon what people make of them.

The presentation and representation of the environment is thus both complex and of practical significance in a world where people are increasingly sensitive to the way in which the actions of individuals and of private and public groups can directly impact both on their lives and on future generations. The following chapters illustrate these important themes and explore their implications.

Presenting and Representing Environments

Presentation, in the sense meant here, is the way in which various forms of information is set out and appears before the audience for which it is intended. However this is accomplished, much is assumed and implied in the content of a presentation with the assumptions and implications seldom carefully spelled out. Not that they are always necessary or expected. This is most often true of the news and entertainment media. For example, the purpose of content in commercial newspapers is essentially to encourage the sale of more newspapers. As Fowler (1991, 222) has pointed out, "News is not a natural phenomenon emerging straight from 'reality', but a product. It is produced by an industry, shaped by the bureaucratic and economic structure of that industry, by the relations between the media and other industries and, importantly, by relations with government and with other political organizations". He goes on (Fowler 1991, 222) to add that in the process of news reporting it is necessary to acknowledge "the power of the structural minutiae of images and words to impose value-laden organisation on news in the process of articulating it". What we gain from news presentations are particular versions of reality that are necessarily selective (Anderson 1997, 107). Environmental events selected for reporting as news are usually those with an element of the dramatic (Anderson 1997, 7) that will catch the readers' attention – tsunamis and floods making people homeless, acid rain decimating forests, or droughts attributed to rising average summer temperatures. Even where all the facts included in the report are accurate, they will be chosen to suit the purpose, with the writing usually aimed at emphasising the dramatic. Part of this is, of course, related to what Hartley calls "news-as-entertainment" (Hartley 1988, 143). Accompanying photographs, or film in the case of television presentation, will be chosen for the same reasons.

Critical analysis of a news presentation will reveal the values incorporated in the text and images, and the degree and leaning of the selectivity. The methods of such critical analysis of news are now very well developed and readily available (Cohen and Young 1982; Hartley 1988). They are largely derived from the well-established field of discourse analysis (Frank 1997; Stillar 1998; Titscher *et al.* 2000; Wood and Kroger 2000; Phillips and Jorgensen 2002) though the majority of news consumers are unlikely to apply such analysis. For them, the information presented may be accepted as true and even as objective, especially if it reinforces their previously existing attitudes and perceptions. The important point is that the content and structure of presentation of the environment is usually the result of subjective value-laden decisions which can have a significant impact in shaping the perception and attitudes of those who use it.

The example of the news used above was chosen because it is part of the everyday world. At least as important is the presentation of the environment in academic research literature. There is an enduring illusion that this is unblemished by selectivity and value-laden decisions. This, of course, cannot be true. What is true is that the most respectable scientific research is expected to adhere to stringent rules of veracity and to produce results that are verifiable, capable of being replicated and repeated. Scientists work within the scientific paradigm. This means that they focus attention on some problems and necessarily exclude others that will not be accommodated by the tools and assumptions provided by their chosen paradigm. Kuhn (1962) pointed out that any one paradigm continues to be adopted as long as it provides a useful guide. When sufficient cases arise with which the paradigm is unable to deal, then the paradigm will either be modified or replaced (Reese 1997). Within the limits of the paradigm there are various opportunities for subjectivity. In the initial stages of any scientific research subjective decisions are made about what particular part of a subject will be investigated. The choice made will be strongly influenced by a range of factors including previous experience, advice from respected researchers in a field, and the human, physical and funding resources available. Choices will also be made as to the particular procedures to be adopted for the research and about the scope of the study. Once the research results are obtained, they are usually submitted for publication. Here, further subjective decisions have to be made as to where the researcher will submit the work for publication and its acceptance by an editor or publisher will also be a subjective decision. For example, academic journals cannot publish every article of academic merit submitted to them and so the editor or the editorial board will decide which to accept. If it is published, the content of the article may make assumptions that are not stated – about the level of knowledge of the reader, about the ideological stance of the researcher and about the technical procedures used. Moreover, the written content is likely to emphasise those elements which the writer will have decided are the most important either in the findings or their interpretation or both.

While reports of scientific results are required to adhere to strict rules about the methods used to produce evidence and the interpretation of the evidence, they also incorporate value judgements and the users of such reports need to be aware of this. It is particularly important since information and understanding about the environment provided by scientific researchers may be used in a variety of ways and for a variety of purposes. One use is to inform decisions on policies and actions about the environment by the public sector or by commercial or non-governmental organisations. For this purpose, it is common for research results to be used to compile a presentation in report form to support a particular conclusion. Generally, this will be either the conclusion the scientist has decided upon or that required by the party commissioning the report. In either case, the presentation can be expected to give preference to evidence that supports a particular viewpoint. This selectivity is facilitated by the fact that environmental knowledge is often incomplete, especially with regard to the processes and the dynamics of an environment. Because of this, the likely result of proposed actions affecting the environment may be in dispute requiring subjective decisions to be made. It is also common for decisions to be taken by people who are not professionally qualified in the field and who will depend for guidance upon those whom they consider to be experts. It is in such cases that the whole process will depend upon selective presentation and value judgements.

Representation of the environment is essentially concerned with the way individuals and organisations perceive and value the environment and choose to convey these perceptions and values to others. The environment can be viewed and represented in a wide variety of different ways. For example, it can be considered as safe or dangerous, it can be represented as having major commercial benefits or as being an obstacle which has to be overcome, it can be represented as having rights independent of humans or as being a resource that should be exploited for human benefit, it can be aesthetically represented as being attractive and desirable or as ugly and repulsive, and so on. Implicit, and sometimes explicit, in these considerations are value judgements so that the way the environment is represented will differ depending upon the underlying ideology and the perceptions and/or purposes of the person or organisation doing the representing (Duncan and Ley 1993, 2). Evidence can be selected from the knowledge available and used to support whichever representation it is desired to make. Frequently, the representation is chosen to try to influence a specified receiving audience. If effective, it may make a vital difference when policies or courses of action are being decided. This is well exemplified in an examination of the news coverage of the environmental risk assessment of a proposed copper mine in Wisconsin (Coleman 1997). There, the representations of each side depended on the two distinct ideological lenses through which the environment was viewed. One was from a scientific, rational and technical worldview and the other a view that valued cultural rationality (Coleman 1997, 483). The former tried to portray proponents of the mine as representing economic progress and advancement through technology. The other view privileged individual, qualitative, experiential and ethical perspectives. The account also suggested that the values guiding the policy of the Department of Natural Resources were essentially those of scientific rationality because they relied on scientists and government officials (Coleman 1997, 486). Such dominance of the values of science and technology are common in environmental policy making. This is largely because they are easier to envisage and deal with (they can be supported by what are perceived to be 'hard' facts) and the views of professional scientists and technologists are often held to be expert and authoritative. This is true for scientists within their paradigm though they are not necessarily or even likely to be informed or authoritative on qualitative, experiential and ethical matters. When more than one cultural group is involved, each with different values, rational science provides only one element of input. In representation as in presentation, the arguments about the significance of subjective value judgements discussed earlier also apply.

Cross-disciplinary Values and the Environment

The environment is a complex and elusive concept and has always been recognised as such. Any one environment is the outcome of the interplay of a wide variety of processes active in the past, many of which are still at work as part of continual change. The mix and strength of these processes vary over space as well as time and it is the patterns and landscapes created over the surface of the earth as a result that are the special concern of the discipline of geography. Nor does the complexity stop there. Environments are experienced, observed and interpreted by different groups and individuals in different ways depending upon their own interests, values and experiences, and these too vary over time. Given this complexity, it is hardly surprising that different disciplines have developed different approaches to describing and understanding the environment and its significance, and that they have devised different ways of presenting and representing it.

For our purposes, a simplified way to understand the different disciplinary approaches is to divide them into two broad categories – those approaching the environment from the standpoint of science and those approaching it from the standpoint of the arts and humanities, including the social sciences. For research, scholarship and study purposes each of these is usually subdivided into subject areas, generally referred to as academic disciplines The term discipline implies, "not just subject matter but also the existence of a 'system of rules' – reproduced through training – for defining that subject matter and the ways in which it is to be studied" (Harriss, undated). These "'rules of conduct', are subject to negotiation and to change, but at any one time they will usually be well defined, principally by the more senior exponents of the discipline, or those in positions of control or command, who thus 'maintain order'" (Harriss, undated). It is important to recognise that, as knowledge and understanding in any one discipline evolve, more and more questions arise which cannot be answered by the existing rules. These rules are then likely to be modified or changed to allow the discipline to progress. Cross-disciplinary approaches to questions are valuable because the resulting cross-fertilisation of ideas and approaches often stimulates better understanding and are a source of valuable suggestions for productive changes of the rules. This is particularly valuable in the environmental field which is not a circumscribed discipline in the way described earlier but which requires inputs from a wide range of established disciplines to advance knowledge and understanding. Cross-disciplinary input is thus not merely desirable but is an essential part of environmental work and decision making.

The standard approach of science is to break down a subject to be examined into component parts to the appropriate level and then investigate and understand each of these components to provide a basis for their re-synthesis to give an understanding of the whole. Critics draw attention to the fact that in too many cases the component parts become the entire focus of attention and that scientists all too often seem to crawl along the limits of their knowledge with a microscope and never see the greater whole. They fail to see the wood for the trees. Others point out that the environment is a whole that is more than the sum of the parts and that the analytical approach is unlikely to provide a basis for appreciating this. The riposte of scientists from particular fields is that intellectually the component parts are of interest, value and significance in their own right and that it is for environmental scientists to tackle the whole. Related to this is the way scientists prefer to present their results. This is normally parsimonious and elegant in the mathematical sense, so scientists may write as much as possible in as few words as possible. Reporting scientific results is seldom seen by non-scientists as 'good writing' or aesthetically pleasing to any but other scientists. The object is clear, precise description to provide knowledge that is irrefutable through methods which others can use to replicate the results. The dangers of this approach are that each sub-science is so far down a particular detailed path that the impact and implications for even nearby subjects can be missed. This is a serious criticism where the environment is concerned (whether the natural or societal environment), since in this field all elements are intricately interrelated in time and space with any change in a component part invariably impacting upon the narrower or wider environmental whole. Geography is an integrating science, focused upon the significance of place as the locale where human and natural environments interacted to create the unique. Geographers have the potential to undertake an even wider integration of disciplines.

The arts and humanities have opted to eschew the scientific approach. They deal for the most part with aesthetics and the conveyance or creation of feelings about a subject or phenomenon. In these disciplines knowledge is seen as personal to the individual. This is based on the recognition that each person is unique with each person interpreting and internalising external stimuli differently. While recognising the possibilities for generalising, academics from the arts and humanities acknowledge that the impact response of any one individual is unpredictable. The knowledge base for what happened in the past cannot be subject to experiment, since each past event occurs in a unique context so that the human environmental context cannot be controlled to replicate it. The result is a dominance of subjectivity and personal interpretation both by the conveyor and the receiver of knowledge. Because aesthetics, perception and emotion have a dominant role in these disciplines, their expression requires them to expand upon a particular theme or subject to achieve the desired outcome for the recipient. Where they exercise parsimony it is often for expediency rather than as a desirable outcome in its own right. For the arts and humanities, the environment is at least as much a matter of morals, emotion and belief as of scientific fact and analysis.

These cross-disciplinary differences, oversimplified here to make the point, are especially important when the environment is being presented or represented. The scientist could be expected to ensure that, as far as is possible, any presentation is firmly based on facts determined by scientific investigation within the current paradigm. From the arts side, a presentation could be expected to direct attention with emotive expression to what people feel about the environment in their experience or through the experiences of others, to involve the value-laden perceptions of the environment and the moral questions involved using ethical arguments. It is possible to identify some overlap between the two approaches, since in any particular case scientists will of necessity have to be selective in the facts they present and will have a value basis for their selection (Kuhn 1962). In the arts and humanities, moral and even aesthetic emotive comments would be expected to have a basis in real world knowledge. The same is true in representation. The emphasis in the arts and humanities will be on the elements which impinge upon the senses of sight, sound or hearing, or which appeal to the individual sensitivities of the receiver. The distinction between the two approaches is not meant to imply that the attributes described are found only in the one or the other. Rigour, parsimony and ethics, for example, are to be found in all disciplines and are a requirement in academic work.

It must be recognised that values lie at the heart of the issues so far described. The cultural filters of the society and/or the individual will impact greatly upon how successful any presentation or representation is. No amount of scientific argument will convince some people that the natural environment is a 'thing' or 'object' to be dealt with unemotionally. No amount of aesthetic/moral argument will convince some people that the natural environment is other than a mechanical/commercial problem/challenge to be dealt with by scientific materialistic logic. For those living somewhere between these two extremes, that is, most of us, both approaches have an impact and can influence our responses to the environment and its challenges. In the following chapters it is easy to distinguish the different disciplinary approaches adopted and the influence of both approaches in their presentation and representation of their subjects. What emerges is that it is not an either/or choice. Different approaches are relevant and important in different circumstances and for different aims. Do not suspend your critical faculties in assessing the arguments and do arrive at your own individual conclusions about the presentation and representation in the environmental challenges presented.

The Evolution of Approaches to the Environment

For all these reasons it is hardly surprising that there have been and will be wide variations in the presentation and representation of any environment. The approaches to dealing with such complexity have evolved over time. The logical-positivist approach that emerged from the Enlightenment of the seventeenth and eighteenth centuries led to an objective science requiring logical structures that were used to identify universal morality and law leading to forms of social organisation based on reason (Harvey 1989). In the twentieth century this emerged as modernism (Gregory 2000). This instrumental rationality was applied to the environment that it was thought could best be understood through the methods of an objective science. It tended to support a societal structure dominated by an elite. This survives even today in "a democratic form of elitism, based on rationality and official expertise (that) sees no need for public participation in solving problems: it knows the answers itself" (Barton 2002, 85). The instrumental rationality of modernism began to be challenged in the 1960s and has gradually given way to postmodernism. At the heart of the new approach is a rejection of metanarratives that allow large scale theoretical interpretations and a means of ordering everything (Smith 1994). Heterogeneity is now recognised as existing and valid and there is an acceptance of the value of alternatives well illustrated by the way collage and eclecticism are so widely accepted in music and the arts (Harvey 1989). In science the shift is illustrated in sub-atomic physics, where laws of physics previously thought to be universal do not seem to apply, and in mathematics, where indeterminacy can be found in fields such as catastrophe and chaos theory and fractal geometry. The philosophical shift is now influencing a different view of the environment resulting in different attitudes and different ways of presenting and representing it. Fixed structures have been replaced by pluralism and alternative cultural values, and governance is seen as having a major role in the ordering of the world (Ley 2000).

It is useful to see contemporary approaches to the presentation and representation of the environment as divided into two fundamental groups – weak environmental theorists seen as technocentrists and strong environmental theorists seen as ecocentrists, with a range of thinking within each (O'Riordan 1996; Williams and Millington 2004). The technocentrists adopt a human-centred position on the relationship between people and nature. This means that people are seen as separate from the natural world that they dominate. Nature is seen as a collection of natural resources available to be used for the benefit of individuals and society. Accompanying this view is an implicit optimism and a confidence that, whatever environmental problems arise, solutions will be found using scientific and technological expertise (Williams and Millington 2004). There is a belief that there will always be a stock of resources sufficient to supply human needs. Accompanying these views is an acceptance that economic growth can continue without a radical change in human demands on the earth, though it is often recognised that resources need to be better used and that there is a need for a more equitable distribution of the associated costs and benefits. Technocentrism is more often associated with big business and government (Castree 2000) and this emphasis is seen in the way tehnocentrists present and represent the environment.

Ecocentrists start from a different position in seeing the earth as having finite resources so that there will be catastrophic results unless demands upon them are much more restricted within what is sustainable. Within the range of attitudes encompassed in this group there is a recognition that nature has biotic rights just as humankind does and that these rights are generally not being respected. They thus reject the human-centred position. It follows that they argue for restraint in the demands made upon nature. They advocate the setting of limits on economic growth which should be seen as a means to an end rather than an end in itself, and a strong emphasis on sustainability within any economic development (Williams and Millington 2004). Ecocentrism is frequently associated with non-governmental organisations (NGOs) and with other, often local, environmental protection groups and is usually readily identifiable in their presentations and representations of the environment.

Cross-cultural Issues

The processes of presentation and representation of the environment obviously presume that judgements have been made as to what aspects of an environment are worth presenting and representing and how these should be presented and represented. There are also judgements to be made about the mode of communication and about the receptivity of the persons and groups to whom the presentations and representations are directed. Implicit in these judgements are values and values are embedded in cultures, distinguishing one culture from another. These judgements are also the products of particular personalities. It is the juxtaposition of culture and personality that leads into an understanding of why presentation and representation are on one side of a coin and mispresentation and misrepresentation are on the other. Thus, in our earlier discussion about cross-disciplinary values, those persons, with their own discrete personalities, who identify with a particular discipline, participate in a broad cultural group sharing common values related to the discipline. It must be recognised, however, that within any discipline there is likely to be a wide array of sub-disciplines, each with active participants who share sub-sets of the broad values that hold together the discipline. Stepping outside of the academic disciplines it is clear that these same generalisations can be applied to social and cultural groups of all kinds, operating in different places and at a variety of geographical scales.

The assertion that cultures have distinctive values introduces an array of concepts such as cultural borrowing, cultural conflict, cultural hierarchy and multiculturalism. In any discussion or investigation about the environment there is always scope for both shared understanding and misunderstanding. One group's interpretation and understanding of what has been presented is often quite different form that of another. Any cross-cultural contact brings this to the fore. What is clear is that the notion that there are national cultures is becoming increasingly useless in an environmental context. Increasing globalisation of environmental challenges – climate change, air and water pollution, destruction of habitats, the production of GM foods – has resulted in cross-national sharing of concern, study and action. While, at a very local level, active groups – local businesses, non-governmental organisations and single interest environmentalists – may illustrate vividly cultural conflict in the ways that they respond to a localised environmental issue – the construction of a village by-pass, the destruction of a copse, the draining of a pond – the values they express are becoming part of a much larger global sharing. There is an important distinction to be drawn between cultural sharing and cultural borrowing. The latter is a deliberate act of searching for inspiration, ideas, practices and artifacts to enhance one's own culture. The presumption is that some cultures are in some way better than others. Thus, in the environmental field, environmental researchers and environmental activists from one culture have much to gain from contact with their counterparts in other cultures. The benefits would be evident in changed knowledge and understanding, attitudinal change and overt behavioural change. In a perfect symmetrical relationship cultural contact would lead to mutual benefits for both cultures.

It is an interesting feature of cultures in an environmental context that while, on the one hand, there is evidence of globalisation of environmental concerns and values, on the other hand, there is evidence of increasing localisation. Cultural groups define in their own ways what they perceive to be their own natural and physical environments. They attribute to aspects of these environments their own meanings, be they religious, social or political. Schama (1995) has comprehensively analysed this in his discussion of the history and culture of forests, rivers, mountains and arcadia. In attributing meanings, cultures express their own worldviews at particular times and for their own purposes. This was well illustrated in a recent exhibition at the Victoria and Albert Museum in London titled *Encounters: the Meeting of Europe and Asia 1500–1800*. The rich array of artifacts – maps, landscapes, portraits and objects – demonstrated the perceptions of Europeans of aspects of Asia and the perceptions of Asians of aspects of Europe at the time of the attempts made by Europeans to open up trade with Asia. Cultural borrowing and cultural conflict stood out but so did cultural mythologising, expressed in the urge to stereotype. Cross-cultural contact serves either to reinforce or shatter such environmental stereotypes and myths. In engaging in cross-cultural studies we need to be alert to achieving a balance between identifying commonalities and differences. It is often difficult to get away from the presumption that our own environmental knowledge, attitudes and behaviour are superior to others. This is true at the very local neighbourhood scale as well as at the global scale.

Social scientists (for example, Ray and Sayer 1999) have referred to the cultural turn that took place in the latter part of the last century as the focus of much work on the humanities and social sciences shifted from the economy to culture. This turn has found expression not only in the topics for study but also in the awareness of scholars of their own cultural identities. As Davies and Gilmartin (2002, 18) assert, "Now the focus is on relationships with society, the representations and meanings within society, and, for some authors, such as Mitchell (2000), a rationale for a radical critique of society, providing a means for creating change, especially in favour of the least advantaged". This is especially evident in, for example, feminism and the environment where there are feminist scholars who form their own disciplinary sub-groups who focus their studies on women's worlds. But it also finds expression in racial groups as well as in groups identified with particular religious, social, political and economic ideologies. A focus on cultural interpretation, understanding and definition of environmental meaning challenges the notion of objectivity, discussed above, and encourages one to be cautious and humble in environmental study. The need to be cautious reminds us of the need to be aware of *who* is presenting and representing the environment and *why.* The humility comes from an awareness that one's own environmental perceptions, understanding and meanings are very limited, rarely are they holistic and universal. Commonly, they are selective, limited and culturally determined.

Themes in the Chapters

This book has been stimulated by an escalating awareness of the importance of the environment in everyday life. There is a growing recognition that what is being presented about the environment and how it is presented and represented are playing an important part in shaping people's attitudes and perceptions. They are also impacting upon decisions about environmental policies and actions that are affecting everyone.

Following the introduction, through an examination of the forestry industry in British Columbia, Roger Hayter illuminates the contested perceptions of natural resource use and the significance of the way the uses of the forest resources are presented and represented. Forests cover much of the huge area of the Province and the forestry industries dominate the provincial economy. Various objections have been made to the terms of the forest extraction licences and their spatial coverage. Environmentalists and significant cultural groups are concerned about the impacts of timber extraction and the lack of sensitivity displayed towards other forest users. The differing perceptions and attitudes revealed in the ways the various groups 'map' the forest resources from their own particular perspectives, derive from their different cultural backgrounds and from conflicting ideas about resource utilisation.

The next chapter demonstrates the changing role of science and of scientists and their importance in environmental matters where major commercial organisations, the public sector and local inhabitants are involved. Using the example of how grizzly bears are dealt with in a part of north west Canada, Doug Clark and Scott Slocombe reveal how easy it is for differently obtained knowledge to have different values for different cultural groups. They show how it is necessary to be sensitive to these differences in deciding what environmental action needs to be taken if agreed and accepted outcomes are to be achieved. The issues they have chosen to highlight revolve around the objectivity of institutionally produced science and the treatment of uncertainties, plus the concerns with the methods of applied science expressed by non-scientists. In this case, the meeting of two different cultures and the way each side presents and represents the environment is being played out in an arena of northern Canada where the federal government and moves towards first nation self-government come face-to-face with the lives of indigenous grizzly bears.

With the growing awareness of contemporary environmental challenges it is easy to forget that there are readily available lessons to be learned from looking at history. Ian Simmons narrates the changes in the moorlands of Britain over the last ten thousand years and contrasts the objective scientific understanding and presentation of the moorland environment with the differing values of users. The objectivity of science is contrasted with the emotional and practical appeal of the moorlands. He draws on scientific and literary and artistic texts to illuminate contrasting representations. The lesson to be drawn is that, whatever our current concerns seem to be, constant environmental change is here to stay, both in the real landscape and in the minds of the those observing and using it. Wayne Davies provides a valuable lesson about the significance of the background and motivation of producers of environmental accounts and the way the accounts are used in interpretations of them. This is very well illustrated in his examination of the narratives of an early Welsh explorer of Canada to expose the value judgements used in critiques of the content and presentation of the Canadian environment. Davies ably demonstrates that exploration narratives can no longer be seen as 'objective' or 'realist' descriptions of the environments observed. This is because they are carefully constructed, revealing as much about the aesthetic preferences of the explorers and of their critics as about the environment explored.

Chapters 6 and 7 show that interaction between the local political scene and the natural environment is a two way process with both sides strongly influenced by the way in which environmental challenges are perceived, presented and represented. This is especially important where the environmental challenge is embedded within large urban areas. There the environment is a significant and influential component of the natural capital of dense urban populations who are in direct contact with, and daily observe, its use and change. Chapter 6 focuses on the environment of Swansea East in south Wales. This area has a history of urban and industrial use reaching back three hundred years. It concentrates on the post-1960 period to show how local attitudes to environmental usage have changed over time, raising issues about the need for direct public participation in the further development of advanced liberal democratic government. Here, increasing education levels among the resident population and the facilitating of ready accessibility to authoritative information via the Internet have shifted the power structure and resulted in differing presentations on environmental issues. In the next chapter Dianne Draper deals with an environment in a much more recent city, focusing on the retained natural landscape along the Bow River Valley in Calgary, Alberta. There the challenge is how to present a successful case for conservation of the landscape based on scientific knowledge and understanding, and to represent its value as natural capital within a local urban political framework where there are competing demands for resources. Commitment to the environment and continual presentation of its local importance are essential requisites in sustaining the valuable natural capital, an increasingly recognised contribution to the quality of life of major cities.

Two very different chapters follow. Tammy Speers, a Canadian researching in Wales, deals specifically with the way the environment is reported in the mass media in Britain. These media now provide most of the information about the environment received by the majority of people in the world. They make especially important contributions to the moulding of attitudes towards various aspects of the environment by the way they present and represent the environment. Understanding of the processes and values involved in the way the media undertake their tasks, as revealed in this chapter, is thus of considerable significance in contemporary society. Robert Newell is an artist with a special interest in painting representations of the natural landscape. He shows how critiques of landscape representation are part of a larger multi-disciplinary negotiation of the meanings of landscape as a whole. Landscape in this context is seen for what it is, an ideological construct and a pictorial conception. Failure to understand this can be considered detrimental to wider understanding and appreciation of the environment. By careful argument, flaws in negative artistic critiques are exposed, showing how they impoverish artistic and environmental appreciation and understanding.

Multiculturalism and the natural environment are the focus of attention in the next two chapters. Colin Williams tackles a subject of growing global importance – the environment as the scene and setting for cultural clashes between minority cultures and the majority culture with which they have to contend. The case study used is the threatened environments of the 25 per cent of the population of Wales who speak Welsh. Fierce debates rage around the policies and actions to ensure these environments are conserved, within which the presentation and representation of their importance by the Welsh speakers themselves and by the other residents of Wales are important influences. The argument is that national identity and the environment, variously defined, are profoundly entwined. It is an argument that Nancy Doubleday looks at from a different angle, by focusing on artistic expression and the multicultural context of the Inuit of the District of Nunavut in northern Canada. There is a growing appreciation that aboriginal representation of the environment stems from a very different basis to that of the dominant Canadian culture. This is of practical importance because, under Inuit leadership, Inuit identity is now being far more strongly articulated than in the past. Their cultural values, derived from and strongly represented in their attitudes towards the environment, are now being much more strongly presented, and especially in the political field. This is resulting in an opening of new avenues of critical interpretation of environmental sustainability as well as possibilities for transcending duality in development in ways that are found elsewhere in the world.

Two final chapters raise the implications of presentation and representation of the environment for environmental education in lifelong learning. Ian Maclachlan, Nancy Bateman and Thomas Johnston demonstrate how different kinds of education can be used effectively to meet the environmental challenges confronting farmers. They do this using the example of cattle farming, the major agricultural land use in Canada. A case example from the Ottawa Valley in eastern Canada shows the practical value of farmers themselves getting involved with such education, with environmental benefits. In a contrasting area, a programme of research at Lethbridge in the Canadian Prairies illustrates the practical importance of the way progressive environmental attitudes are presented for lifelong learning to achieve environmental improvement. Michael Williams takes the global view in presenting a framework for understanding the importance of environmental education and lifelong learning. In particular, he identifies the constraints and challenges confronting educators as they determine how best to present and represent environmental issues to their students. Educators must choose in a very transparent way the environments they wish to present and represent to learners at different stages in their intellectual development. They must also choose and justify how best to present and represent the environments they have selected. The inadequacy of passive learning in this context is emphasised and the shift from environmental understanding to environmental action is highlighted.

The chapters blend case examples with theoretical underpinnings in a way that demonstrates the vital significance of presentation and representation of the environment in the practical world of policy making and of actions, decisions and implementation. Focused on Wales and Canada, they raise continuing issues applicable across the world. They are thought provoking and have implications for the way that we confront contemporary and future environmental challenges to the quality of life of us all.

References

Anderson, A. (1997). *Media, Culture and the Environment.* London: UCL Press.

Barton, B. (2002). Underlying concepts and theoretical issues in public participation in resources development. In D. N. Zillman, A R. Lucas, and G. Pring (Eds.) *Human Rights in Natural Resource Development: Sustainable development of mineral and energy resources.* Oxford: Oxford University Press.

Castree, N. (2000). Environmentalism. In R. J. Johnston, D. Gregory, G. Pratt and M. Watts (Eds.) (4th Edn). *The Dictionary of Human Geography.* Oxford: Blackwell.

Cohen, S. and Young, J. (Eds.) (2nd Edn) (1982). *The Manufacture of News: Deviance, social problems and the mass media.* London: Constable.

Coleman, C. (1997). Science, technology and risk coverage of a community conflict. In D. Berkowitz (Ed.) *Social Meanings of News: A text reader.* London: Sage.

Davies, W. and Gilmartin, M. (2002). Geography as a cultural field. In R. Gerber and M. Williams (Eds.) *Geography, Culture and Education.* Dordrecht: Kluwer.

Duncan, J. and Ley, D. (Eds.) (1993). *Place/Culture/Representation.* London: Routledge

Fowler, R. (1991). *Language in the News: Discourse and ideology in the press.* London: Routledge.

Frank, M. (1997). *The Subject and the Text: Essays in literary theory and philosophy.* Cambridge: Cambridge University Press.

Gregory, D. (2000). Modernism. In R. J. Johnston, D. Gregory, G. Pratt and M. Watts (Eds.) (4th Edn) *The Dictionary of Human Geography.* Oxford: Blackwell.

Harriss, J. (undated). The case for cross-disciplinary approaches in international development, www.lse.ac.uk/collections/DESTN/pdf/wp23.pdf (accessed November 2004).

Hartley, J. (1988). *Understanding News.* London: Routledge.

Harvey, D. (1989). *The Condition of Postmodernity*. Oxford: Blackwell.

Kuhn, T. S. (1962). *The Structure of Scientific Revolutions*. Chicago: Chicago University Press.

Ley, D. (2000). Postmodernism. In R. J. Johnston, D. Gregory, G. Pratt and M. Watts (Eds.) (4th Edn) *The Dictionary of Human Geography*. Oxford: Blackwell.

Mitchell, D. (2000). *Cultural Geography: A critical introduction*. Oxford: Blackwell.

O'Riordan, T. (1996). Environmentalism on the move. In I. Douglas, R. Huggett and M. Robinson, (Eds.) *Companion Encyclopaedia of Geography*. London: Routledge.

Phillips. L. and Jorgensen, M. W. (2002). *Discourse Analysis as Theory and Method*. London: Sage.

Ray, I. and Sayer, A. (1999). *Culture and Economy after the Cultural Turn*. London: Sage.

Reese, S. D (1997). The news paradigm and the ideology of objectivity: A socialist at the Wall Street Journal. In D. Berkowitz (Ed.) *Social Meanings of News: A text reader*. London: Sage.

Schama, S. (1995). *Landscape and Memory*. London: Fontana Press.

Smith, D. M. (1994). *Geography and Social Justice*. Oxford: Blackwell.

Stillar, G. F. (1998). *Analysing Everyday Texts: Discourse, rhetoric and social perspectives*. London: Sage.

Titscher, S., Wodak, R. and Vetter, E. (2000). *Methods of Text and Discourse Analysis*. London: Sage.

Williams, C. C. and Millington, A. C. (2004). The diverse and contested meanings of sustainable development, *The Geographical Journal*, 170, 99–104.

Wood, L. A. and Kroger, R. O. (2000). *Doing Discourse Analysis: Methods for studying action in talk and text*. London: Sage.

2. ENVIRONMENTALISM *QUA* ENVIRONMENTAL NON-GOVERNMENT ORGANISATIONS AND THE CONTESTED REMAPPING OF BRITISH COLUMBIA'S FORESTS

ROGER HAYTER

Introduction

British Columbia's (BC's) vast forest economy has been in the eye of environmental storms for over three decades. Greenpeace, whose evolution captures the growing strength of professional environmental non-government organisations (ENGOs) around the world, was founded in Vancouver in 1971 by a handful of protesters opposed to atomic bomb testing (at that time in Alaska) and the industrial destruction of forests. Although they vary in philosophy and tactics, these ENGOs collectively argue for restraint in the demands made upon nature, with a strong emphasis upon sustainability within any economic development (see Chapter 1). In presenting their position they privilege emotive, aesthetic, moral and cultural values above economic growth and development. They represent the environment as natural and precious, needing stewardship and nurturing as opposed to destructive exploitation. It is little wonder then that, in regions where natural resource based industries dominate, they clash with the commercial interests seeking to maximise the return from their investments. Governments too, where they are seen to be putting the economic prosperity of their constituents above environmental sustainability, are seen as the enemy. In BC the ENGOs strongly promote the ecological values of the environment, and use a variety of allies in the presentation of their case. They have been remarkably successful in fighting their corner, though as this study reveals, even where they win their case the outcome is not necessarily a success for the values that they hold dear.

With respect to forestry, sporadic, localised opposition to logging occurred during the 1970s. The environmental clock on BC's forest economy was ticking faster. In the context of broadly based social protests in 1983 directed against 'right wing' provincial government policies and a deep recession that was especially devastating to resource sector firms, communities and employees, ENGO opposition to industrial logging escalated in terms of organisation, tactics and motivation. This escalation became especially apparent in the Clayoquot Sound protests of 1993 when international ENGOs organised thousands of demonstrators seeking to preserve the area's ecological

G. Humphrys and M. Williams (eds.), Presenting and Representing Environments, 17-32.

values 'for the globe'. ENGO motivations were now committed to achieving their representation of the environment through remapping British Columbia's forests from industrial use towards conservation. These goals were spearheaded by the promotion of the so-called Great Bear Rainforest and the Yellowstone to Yukon (Y2Y) corridor.

ENGO opposition to industrial forestry in BC has been reinforced by the powerful, related criticisms of aboriginal and US-based trade interests. The Coalition for Fair Canadian Lumber Imports (CFCLI) has complained since 1981 that BC forest policy subsidises its lumber exports to the US and has effectively lobbied for restrictions on the import of Canadian lumber since 1986. In 2001, a remarkably punitive tariff of 27 per cent was imposed on BC's lumber exports to the US. Also over the past two decades, BC's Aboriginal Peoples, or First Nations, have re-invigorated their demands for treaties and opposed industrial use of forests on land subject to treaty claims. Stimulated by shared criticisms, ENGOs have diligently sought to forge alliances with Aboriginal Peoples and the CFCLI to undermine prevailing industrial forest policy. In response, the provincial government, which controls over 90 per cent of forest land in BC, has strenuously sought solutions to these 'wars in the woods', including by a battery of environmentally inspired legislation to meet a wider range of ecological and social goals and a sustainable industry. This chapter reflects on the troubled relationships and political economy between ENGOs, with their ecocentrist representation of the environment, and provincial government forest policies that seek to restructure BC's forest economy on an environmentally and economically sustainable basis. The chapter is especially concerned with the implications of provincial initiatives and ENGO activity for the restructuring of industry from its established large-scale, cost-minimising commodity orientation towards higher value, more diversified activities that rely less on first or old growth timber. Restructuring trends in this regard, however, remain ambiguous. A concern raised by this chapter is that ENGO activism that has demanded a profound rethinking of provincial forest policy has also helped undermine industry confidence in contemplating investment in a more innovative sustainable future.

The discussion unfolds in four main steps. The first part provides a brief conceptual context by noting the widespread occurrence of contested resource peripheries in the present global economy and the central role of ENGOs in these conflicts. BC's forest economy is then presented as an exemplar of a contested resource periphery and an especially important target for ENGOs. The third part summarises the goals and tactics of ENGOs in BC. The fourth part relates ENGO behaviour to provincial government policy initiatives of successive governments, especially from 1988 to 2003, in meeting an environmental agenda for the forest economy. This discussion suggests that, while ENGOs can claim much credit for the implementation of this agenda, ENGO behaviour has also compromised the restructuring of BC's forest economy towards value maximisation (and away from commodity production based on cost minimisation). The discussion is inevitably highly generalised but draws on a considerable recent social science literature on BC's forest economy, including detailed analyses of restructuring processes (Hayter 2000; 2003), government forest policy (Cashore *et al.* 2001) and environmentalism (Wilson 1998; Stansbury 2000).

Restructuring and Contested Resource Peripheries

In contemporary debates, the restructuring of the global economy over the last several decades, driven by technical change, new market forces and increasingly severe recessions, is often summarised as involving a shift from Fordist to post-Fordist production systems. In general terms, this restructuring implies a search for economies of scope as well as scale as more flexible firms, technologies and firms replace, or at least modify, the mass production cultures of Fordism. Within economic geography, the idea of restructuring was pioneered primarily in relation to the deindustrialisation of old industrial agglomerations and the emergence of so-called 'new economic spaces' that were dominated by secondary manufacturing or service activities. Resource peripheries were largely theoretically marginal to this discussion (Hayter 2003). Yet around the world, economic development in a large number of resource peripheries has become a highly contested process in a wide variety of 'resource wars' (Ecologist 2002; Hayter *et al.* 2003). Moreover, these conflicts are created by a complex interaction of institutional forces that are not found in old (or new) industrial agglomerations. Thus resource peripheries, as *industrial* regions, have been deeply affected by the restructuring processes of post-Fordism that are spearheaded by the imperatives of flexibility. Resource peripheries must confront and negotiate these imperatives, however, from vulnerable spaces on geographic margins that are rooted in remoteness, external control, specialisation and extreme export dependence. Consequently, resource peripheries have been vulnerable to trade politics and the tendency of rich countries to impose restrictions on resource imports, according to the wishes of domestic lobbies.

Resource peripheries are also *resource* regions, deeply affected by the peculiar characteristics of resource cycle dynamics, and by cultural defining of resources that are valued both as inputs to industrialisation and for wide ranging ecological and cultural, including aesthetic and spiritual, ('non-industrial') benefits. Moreover, the resource cycle hypothesis argues that the exploitation of both renewable and non-renewable resources means that as resources deplete they become more costly to industry as the best and most accessible are used first. Depletion also threatens or even destroys their non-industrial values. During post-Fordism, however, public attitudes have fundamentally changed in favour of the non-industrial values of resources, even as industrial operations have faced increasing cost disadvantages. Indeed, 'environmentalism', broadly conceived as institutional commitments that seek to reduce human impacts on the natural environment, has especially profound consequences for resource peripheries.

Indeed, the radical innovations in information and communication technologies underlying post-Fordism that have deepened the global integration of capital have also empowered ENGOs to influence public opinion (and policy) and resist this deepening. The tactical strengths of ENGOs relate to their communication abilities to shape public opinion, lobby governments, pressure corporations, set agendas and organise campaigns against industrialisation in the world's most remote places. Information technologies that have compromised territorial governance have proven ideally suited to the proliferation of boundary-spanning ENGOs whose global reach matches that of their principal adversaries the multi-national corporations (MNCs). Indeed, ENGOs are able to target the remotest resource peripheries that are sites of resource depletion and where the destruction of bio-diversity has global significance in real and symbolic terms. Thus, high profile, belligerent ENGO campaigns have featured anti-mining, anti-fishing (including anti-whaling and anti-sealing), anti-oil drilling and anti-logging activities using a wide range of tactics.

Contested resource peripheries also feature resistance to industrialisation by surviving aboriginal populations who have their particular cultural interpretations of resource values, broadly motivated by a sense of self-identity, aboriginal rights, and self-control over development. ENGO strategy is to ally themselves as far as possible with Aboriginal Peoples in an effort to portray similar 'caring' views about the environment. If such alliances have been widespread, relationships between ENGOs and aboriginal groups have also often been fractious in both developing and advanced countries.

According to the contested resource periphery hypothesis, the conflicts that are occurring around the world among resource peripheries are a systemic, defining feature of 'globalisation' driven by unique clashes of 'institutional voices' that are demanding that resources be re-allocated and remapped according to their interests (Hayter 2003; Hayter *et al.* 2003). In this view, the resource wars of contemporary globalisation are not coincidental but driven by the scale and intensity of Fordism's resource cycles. More specifically, changing social attitudes are demanding that industrial exploitation and restructuring take into account environmental imperatives. The rise of ENGOs has both reflected and shaped changing social attitudes to the environment (O'Riordan 1996). As professionalised advocates of environmental values, ENGO activities have been controversial and have imposed new forms of global-local dynamics on local development (Hayter and Soyez 1996).

In resource peripheries, ENGOs are global and local institutions (Hayter and Soyez 1996). As local institutions, ENGOs are democratic voices, urging for greater local control over resources, greater transparency in resource allocation, and wider (stakeholder) participation in resource use. Indeed, ENGOs began as local opposition to the global force of industrial destruction, and, for Paehlke (1996), enlightened environmentalism is fundamentally democratic in that it depends on societies that celebrate individual initiative and because it deliberately seeks to engage the public in the environmental agenda. As global institutions, Princen (1994) argues that ENGOs provide 'countervailing power' to MNCs (and irresponsible governments) to ensure transparency in informing the public of all the implications of industrialisation and seeking to ensure that action is taken at appropriate 'bio-geographical scales'. He also emphasises that ENGOs gain social legitimacy by their sole focus on environmental issues.

As global institutions, however, ENGOs impose values on resource peripheries, regardless of local priorities or preferences. As Taylor (1996) argues, ENGOs, empowered by their single-issue focus and global scope, are prone to authoritarian thought. The scientific basis for resource scarcity issues is rarely straightforward (even among scientists) and ENGOs can communicate misleading and ambiguous messages to further their goals. There is also a deep concern, even within developed countries, that ENGOs ideologically reflect the agenda of urban elites. From a Pacific Northwest perspective, for example, Arnold (1999) notes that ENGO agendas are increasingly controlled by urban-based endowment funds and Lee (1994: 228) argues ENGOs "using guilt, shame and ridicule to control people", often supported by federal bureaucracies and trust funds, are creating a "cultural upheaval" in the region as they seek to monopolise policy in federal forest lands. The recent tactic of ENGOs to rename and 'claim' resource regions as conservation areas graphically reflects the dissonance between urban and rural values.

In this regard, the targeting of resource peripheries may reflect ENGO public relations' concerns as well as over-exploitation itself. Despite a few attacks on sports utility vehicle owners, ENGO belligerence has been less evident in congested urban areas where there are large populations, than in sparsely populated peripheries. ENGOs, for example, have not mounted anti-commuting blockades in New York or London to complement anti-logging blockades. Moreover, to justify operations (and funding), ENGOs may target resource regions where success is more likely rather than necessarily most appropriate. It has been further suggested that ENGO targeting of resource regions, firms and projects may fail to grasp wider implications, generating unintended negative impacts on the environment. Specifically with reference to BC, for example, Moore (2000) a founder of Greenpeace, argues that unrelenting ENGO attacks on forestry are misinformed and increase environmental problems as the ecological benefits of young, growing forests are lost and switches are made to more environmentally damaging substitute products.

Target BC: Forest Restructuring *Qua* Remapping in an Environmental Perspective

BC is a huge area of 95 million hectares, about twice the size of Germany, but with less than four million people. Forests occupy about 63 per cent of BC and commercial forest land amounts to 23 million hectares (over 24 per cent of the total) and in 2000 log production was 77 million cubic metres. Over 90 per cent of BC's forests are publicly owned. About one-third of the contemporary timber harvest is from the coastal region, historically the most important until 1970 since when production throughout the interior has expanded considerably (Figure 2.1). In the cool temperate rainforests of the coast, balsam, hemlock, western red cedar and Douglas Fir dominate the harvest while lodgepole pine, spruce and true firs are the predominant interior species. The share of second growth forests is increasing but significant quantities of old growth timber remain. The 1996 log harvest of over 75 thousand cubic meters mostly supplied large-scale production in support of a forest sector, comprising one hundred thousand direct jobs and manufactured shipments worth Can$16.5 billion. Forestry remains BC's most important industrial sector and in the 1990s about 90 per cent of the sector's output was exported.

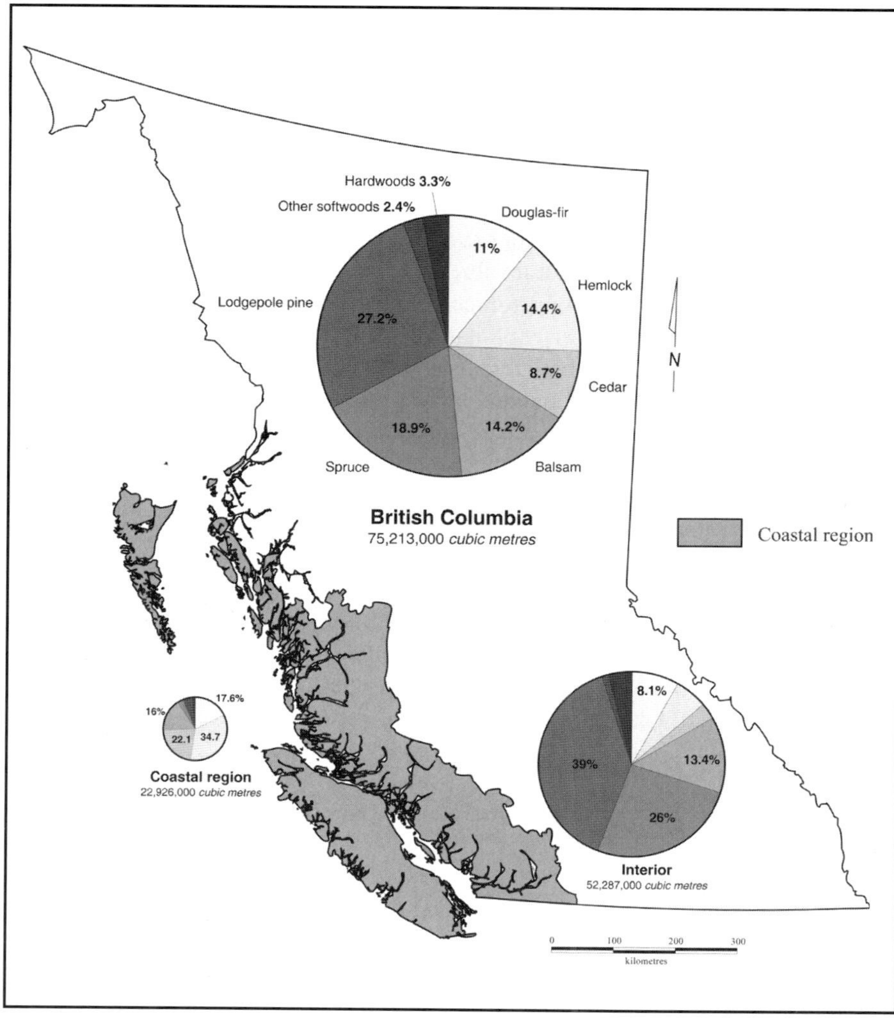

Figure 2.1 *Log production in British Columbia by species and region, 1996.*

ENGO protests of logging were already in full swing before Greenpeace identified BC as the priority target of its forestry campaigns. That BC should be such a target is not surprising. The province contains a huge, ecologically rich, diversified forest resource, large areas of old growth are intact and, bearing in mind the relatively low population of about three and a half million people, there is considerable potential for the remapping of forests to meet environmental criteria. The importance of public control and export dependence also implies vulnerability to ENGO targeting. Moreover, industrial forestry in BC is primarily developed around a mass production model largely organised by large corporations, including MNCs.

The Historical Perspective

The exploitation of resource wealth, especially forest resources, has keyed BC's incorporation into the global economy over the past one hundred and twenty years. Initially, logging literally meant mining the vast old growth forest wealth of the province and speculation and forest destruction, especially in coastal BC, became rampant until the provincial government stopped sales and leasing of timber rights in 1913, following the advice of a Royal Commission on Forestry. Following another (Sloan) Royal Commission on Forestry, the Forest Act of 1947 sought to establish a policy framework to encourage stable and sustainable economic development throughout the province by privileging the role of large corporations, foreign or Canadian. Basically, this policy granted giant timber tenures over lengthy (and renewable) periods of time at relatively low cost to corporations in exchange for major investments in processing facilities within the area of the leases. Socially, such investments were seen as long term, community-building commitments, reinforced by principles of sustained yield forestry. In this latter regard, the Forestry Act of 1947 proposal for sustained yield forestry was principally through the mechanism of the annual allowable cut (AAC) which the government established for each lease. In principle, AACs were set to allow forest renewal to occur in time for a second cut. By requiring firms to cut their AAC (over a five-year period) or forfeit their timber rights, the government sought to achieve economic development of the forestry resources that was sustainable.

In practice, estimating AACs was judgmental, secretive and involved data provided by firms while, in the absence of evidence to the contrary, for all intents and purposes forest renewal relied on natural regeneration. If the process of setting AACs was flawed, natural regeneration proved an excessively optimistic belief. More fundamentally, the Forest Act of 1947 enshrined the equation between forest policy and industry policy. The value of a tree was its value as dimension lumber, or toilet paper as ENGOs now emphasise, and virtually the entire forest resource seemed scheduled for the logger's axe and saw. Indeed, the prevailing logging philosophy advised that BC's vast natural, mature ('old growth') forests should be cut before they became 'decadent', thereby losing further (economic) value through fire, disease and pests, while new, faster growing forests would be ecologically beneficial. Indeed, this view continues to be important (Percy 1986), while recent, ongoing damage by fire and pests (especially the pine beetle) has probably re-enhanced its credence.

Environmental Criticism and ENGOs

The environmental criticism of BC's forest policy was that it discounted the ('non-industrial') aesthetic, spiritual and ecological values of forests, specifically old growth forests. In this view, old growth trees are not 'decadent' but part of natural evolutionary cycles that do not coincide with harvesting regimes. In the early 1970s, symbolised by the early activities of Greenpeace, environmental criticisms of BC's forest policy became more vocal. Mounting criticisms reflected the remarkable growth of BC's forest industries, as anticipated by the Forest Act of 1947, and the scale of logging activity, as well as growing global concerns for 'spaceship earth'. Moreover, by the 1970s rates of growth in the forest sector began to fluctuate and employment levels peaked. The severe recession of 1981-1985 was especially devastating to BC's economy as investment, output levels, profits and jobs all plunged. On the other hand, for ENGOs the recession confirmed that BC's forest economy was not sustainable, economically or environmentally.

To a considerable degree environmental campaigns to discredit BC's forest practices have focused on 'clear-cutting', a method of log harvesting in which all trees are logged within designated cut-blocks. BC's forest policy has long favoured clear-cutting for economic, worker safety and environmental reasons while the practice of 'continuous clear-cutting' meant that logging occurred in adjacent clear-cuts one year after another, thus creating very substantial logged over areas. For ENGOs, clear-cutting is unnatural and results in all kinds of ecological damage, as well as being visually displeasing. Moreover, ENGOs claimed that such logging, continually declining as a job generator, was a threat to growing tourism, recreational and fishing activities.

The CFCLI and contemporary aboriginal opposition to BC's forestry practices was also stimulated by the early 1980s recession. In the US, as in BC, the recession had been equally devastating to lumber producers but the latter could chose to emphasise Canadian imports as the source of their problems. In particular, the CFCLI claimed that while they had to pay 'market prices' for their logs in Canada, and especially BC, the government subsidised the forest industry by charging a low 'stumpage' (tax) on logs harvested on public (leased) land. Meanwhile, Aboriginal Peoples became fearful of resource devastation in the absence of treaty rights and readily joined in protests opposing BC's forest policy.

For BC's forest industries, the severity of the recession underlined a need to restructure, especially in the longer established coastal industry where facilities were older and labour-union relations particularly fractious. Indeed, with gathering momentum industry emphasised modernisation, labour flexibility, trade with Japan, and the manufacture of 'higher value', that is more processed, products. By the early 1990s, government reports recognised a so-called 'value-added' industry, a trend also supported by environmental arguments, in principle at least.

ENGO Behaviour

Since the foundation of Greenpeace, environmentalism within the BC forest economy has become dominated by professional, permanently staffed ENGOs with organisational structures and strategies and global connections to ENGOs elsewhere. Greenpeace and other big ENGOs are multinational organisations themselves; in Greenpeace's case its global head-office is in Amsterdam. In addition to size and scope, ENGOs differ considerably according to beliefs and tactics (Wilson 1998). In general, however, ENGOs have become highly visible, active and controversial participants in BC forestry debates. Bearing in mind their differences, several key features of ENGO opposition to the established culture of mass production need to be noted.

Fundamentally, ENGO behaviour in BC's forest economy, as in many (but not all) parts of the world, is rooted in highly adversarial and uncompromising attitudes. In BC's forest sector labour relations have a long adversarial history but, unlike unions, ENGOs are not 'part' of firms, and they do not share common interests in industrial processes, market viability or profit sharing. Logging blockades were angry confrontations whose immediate goals were literally to stop logging of particular areas, regardless of effects on union workers as well as management and owners. To be sure, ENGO angst has often not been resolved to their liking and compromise decisions have been accepted. It may be that in recent years ENGO criticism has become muted. Nevertheless, there are oppositional, uncompromising roots and cores to ENGO behaviour in BC and ENGOs continue to portray themselves as independent, 'third-party' invigilators of environmental practice.

In BC, where most forests are public owned, ENGO criticisms have inevitably been directed towards government forest policy as well as big business. Moreover, in BC where the provincial political system is highly polarised, both 'right wing' and, less expectedly, 'left wing' governments have felt the brunt of ENGO concern. Thus the explosion of ENGO protests in the mid-1980s readily linked bad business practice with the policies of the right wing Social Credit (Socreds) that were closely associated with the pro-big business policies stemming from the Forest Act of 1947. Yet, when the left wing New Democratic Party (NDP) was elected in 1991, following a short honeymoon, ENGOs proved to be relentless critics of the new government's comprehensive package of environmentally motivated forest policies. Indeed, ENGOs ultimately campaigned against the NDP (in favour of a Green Party), helping to return a right wing (Liberal) government with an overwhelming majority (and no seats for the Greens). ENGOs are hard to please!

Another striking feature of ENGO behaviour in BC's forest economy is its escalation in terms of form, size, nature and organisation of protests, and goals. As illustrated by the formation of the Friends of Clayoquot Sound in 1979 to protest logging on Meares Island, ENGO opposition to the forest industry began as *localised* logging blockades, transparently organised by *local* people who often sought legal support to preserve *locally* prized forests. By the mid-1980s, ENGOs threatened and pursued valley-by-valley conflicts in which the Stein, Walbran and Elaho, as well as protests in Clayoquot Sound and the Queen Charlottes were especially well publicised events (Figure 2.2). Court injunctions were lauded if favourable, ignored if not, and even criminal behavior such as tree spiking occurred. By the time of the Clayoquot Sound protests of 1990-1993, when over a thousand people were arrested, ENGO opposition to the forest industry was orchestrated internationally, organised principally by Greenpeace International and its affiliates, along with the US-based Rainforest Action Network (RAN) and Natural Resources Defence Council (NRDC). Logging blockades were complimented by globally based market boycott campaigns, sources of financing had shifted towards Trust Funds based in the world's leading cities, and

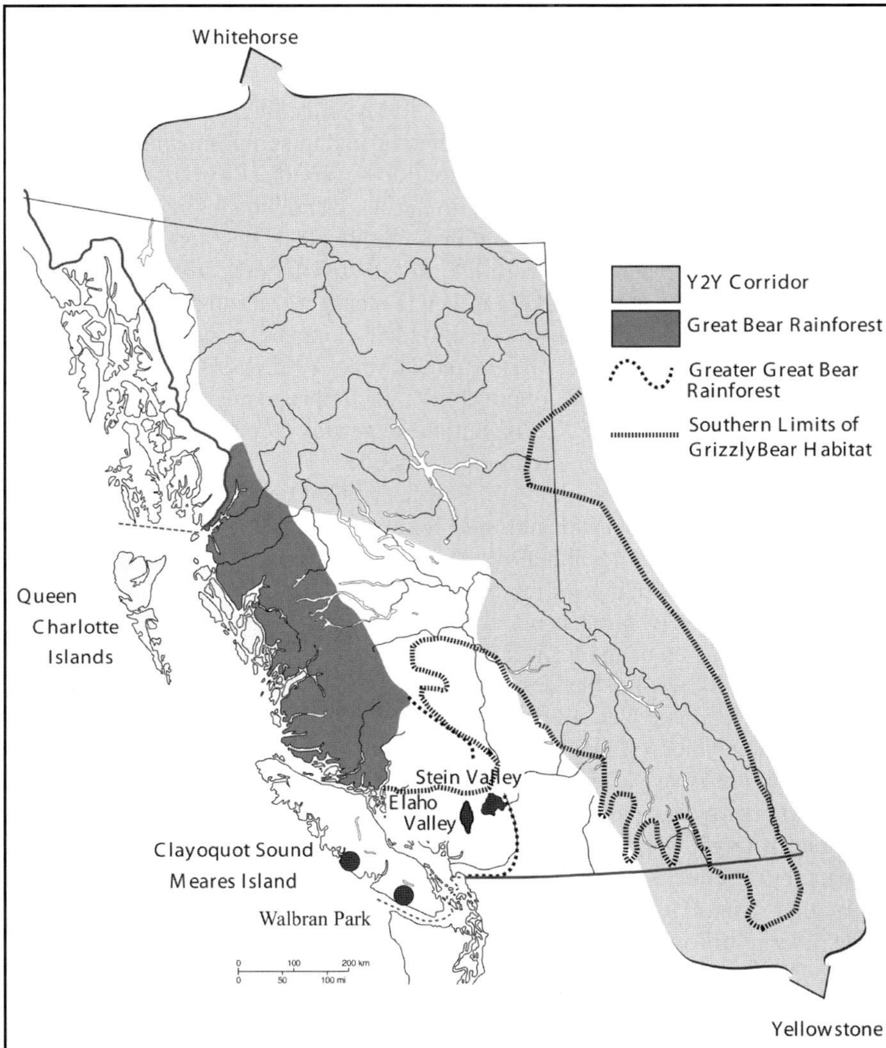

Figure 2.2 *British Columbia's contested space: environmental imperatives.*

protests were publicised by skilled, even manipulative, use of the media across North America and Europe.

Following the Clayoquot protests of the early 1990s, the escalation of ENGO motives culminated in the form of remapping proposals for vast areas of BC (Figure 2.2). In particular, ENGOs launched an ambitious global campaign to preserve old growth forests in BC's coastal region, including much of Vancouver Island, which they would like to rename as the Great Bear Rainforest, after the white *kermode* bear (a genetic variation of the black bear) that lives there. Global bio-diversity is also the basis for ENGO proposals to preserve a vast swathe of interior BC, known as the Y2Y (Yellowstone to Yukon corridor). Recently, ENGOs provided instant support for a recent study that suggested the need for much of northern BC, overlapping considerably with the Y2Y proposal, be left or returned to the wild to allow grizzly bears to survive (Read 2004).

Within the context of escalating, adversarial tactics, it is worth noting that Greenpeace not only made a decision to target BC as a campaign priority but within BC it suggested a strategic focus on MacMillan Bloedel (MB). This priority made sense as MB was the largest, highest profile forest product corporation in BC, the biggest tenure holder, and all its tenures were in coastal BC. MB was also headquartered in BC and, among forestry firms, a leading innovator, until its acquisition by the much bigger, US-based Weyerhaeuser Corporation in 1999. Given the emphasis on blockading MB logging, boycotting MB products, disrupting MB's annual meetings and the general shaming of MB's behaviour, ENGO opposition to MB's acquisition was somewhat ironic.

Another important feature of the escalating, adversarial tactics of ENGOs is their zeal in organising alliances in combating BC's forest policy. From their beginnings as independent local initiatives, ENGOs have been especially effective in organising and coordinating themselves within BC and around the globe, as illustrated by the orchestration of the Clayoquot Sound protests in the early 1990s by global and local ENGOs (Stansbury 2000). ENGOs have also successfully hidden from public view their differences. Simultaneously, ENGOs have built bridges with the CFCLI and Aboriginal Peoples. Both the CFCLI and Aboriginal Peoples, as well as ENGOs, have escalated their demands on government for reform of forest policy (see Hayter 2003).

Moreover, each group has become mutually supportive in this escalation. ENGOs and the CFCLI share deep antagonism towards BC's forest industry and have little concern for its viability. The very different ideologies, and proposed solutions, between the CFCLI and ENGOs have therefore not been of practical concern to them. Thus ENGOs have strongly supported the CFCLI claims that BC's stumpage is a subsidy and have further claimed that Canada's relatively weaker environmental policies are another form of subsidy that undermines the US industry. ENGOs have typically raised public criticism and published damaging reports on BC's forest practices to coincide with major trade negotiations over the softwood lumber dispute. In return, the CFCLI critiques of BC's forest policy, buttressed by punitive measures, have provided a 'business legitimacy' to ENGO positions. However, opportunistic CFCLI and ENGO representatives have stood side by side on public forums in publicising their criticisms of BC's forest policy. ENGOs have also supported treaty claims in association with the portrayal of Aboriginal Peoples as living in sustainable harmony with the environment. There have been public tensions in this association but ENGOs have been tireless workers in mending fences and always supporting treaty claims. Should the signing of treaties maintain large-scale forestry, albeit controlled by aboriginal groups, tensions may reappear.

ENGOs and the forestry labour unions, however, have never really accommodated one another. During the Clayoquot Sound protests for example, the unions organised an even bigger counter-demonstration. Yet, labour's role is long forgotten in this episode while the ENGO demonstration has become part of environmental (and Canadian) history. For a variety of reasons, the influence of once extremely powerful forest unions has declined since the recession of the early 1980s. Their decline along with the rising influence of ENGOs is not just coincidental. ENGOs have been relentless critics of mass production – the basis for the union movement – and have not infrequently expressed concern about the excessive wages of forestry workers in BC.

The Environmental Re-regulation of Forest Policy 1988–2003

What about actual forest policy? Has this policy become more environmentally friendly and consistent with ENGO goals? These questions are not easy to answer. Wilson's (1998) comprehensive, sympathetic opus to environmentalism within BC concludes with a rather disappointing view of ENGO effectiveness and profound frustration over the lack of democracy in (forestry) decision making. Cashore *et al.*'s (2001) sympathetic analysis of NDP forest policy initiatives also concludes that the environmental implications of these policies, however well-meaning, are disappointing. As a business scholar, on the other hand, Stansbury (2000) documents the powerful effects of ENGO belligerence on firms, communities and people's lives. I suggest that, even if results have not met their expectations, ENGOs have exerted far-reaching impacts on BC's forest policies (and practices) and that there are significant unintended effects that are not fully appreciated (Hayter 2000; 2003).

In legislative terms, the provincial government response to environmental concerns has been considerable, most notably by the (left wing) NDP between 1991 and 2001. The previous Socred administration had ended in disarray and its inability to cope with the war in the woods was a major contributor to its downfall. Nevertheless, in 1988 the Socreds did increase stumpage (taxes) on crown timber in an effort to meet both CFCLI and ENGO demands that BC's forests were under-valued. In the event the Socreds were annihilated at the polls and the NDP promised more comprehensive plans to bring peace to the woods. This promise was realised through a remarkable battery of policies that collectively constitute an attempt to 're-regulate' BC's forest economy. There were three main drivers to the re-regulation, notably major changes to stumpage policy, the Forest Code and an expansion of the provincial park system including the (virtual) preservation of Clayoquot Sound.

Thus, stumpage was progressively increased on a yearly escalating formula and minimum levels of overall stumpage payments were established. By the mid-1990s, BC's stumpage costs were among the highest, if not the highest in the world. The new (1995) Forest Practices Code drastically changed forestry. In particular, logging had to meet environmental concerns, for example, by significant reductions in the size of clear cuts, logging bans near streams, and the discontinuance of continuous clear-cutting. In addition, log harvesting plans had to meet new aesthetic and ecological criteria and had to be submitted to an entirely transparent process requiring extensive, time consuming public input, including specific aboriginal input. The park system was more than doubled to 13 per cent of the provincial land base and a special decision making process for Clayoquot Sound was introduced, initially known as the Clayoquot Sound Compromise. In addition to these policies, the NDP announced timber supply reviews (with a view to downsizing AACs), rigorous pollution laws were passed, more timber was allocated to small firms, and new regional planning procedures were introduced along the lines urged by environmentalists. Moreover, to appease aboriginal demands, the government recognised the 'inherent rights' of Aboriginal Peoples and urged the Federal Government to cooperate in a new Treaty Process. Tree planting that had been increased in the 1970s was also significantly expanded; typically BC now plants more trees than it cuts.

Remarkably, ENGOs were largely unimpressed by these initiatives. In many ways, the ongoing lightning rod of these criticisms was the provincial government's continued commitment to clear-cutting, despite the changes introduced by the Code. ENGOs variously argued that either clear cutting should stop in principle or that the new regulations were not effectively implemented. In fact, such has been the vehemence of ENGO opposition that industry itself has begun to implement new practices, notably variable retention logging. In this system, clumps of trees are left in places that make most sense ecologically (for example bird nesting, soil protection). ENGOs also objected to the Clayoquot Compromise that proposed greatly reduced logging rights for the area. In this case ENGO unwillingness to compromise paid off in the creation of the UNESCO Biosphere Reserve in 1999 and the virtual elimination of logging (Hayter and Soyez 1996). Subsequently, ENGO persistence over the Great Bear Rainforest paid off in a 2001 agreement in principle with the NDP, in which logging was again significantly reduced in an area of the central coast that was part of the environmental land claim. Provoked by ENGO pressure, industry and the government are taking forest certification seriously. Here again, however, disagreement exists over who should certify as ENGOs prefer the Forest Steward Council (FSC), a 'third-party' ENGO based in Mexico, even though it is struggling to establish standards for BC. With regard to the remarkable expansion of the provincial park system, ENGOs expressed some satisfaction but saw such expansion as a 'first step' as they escalated their demands for further areas of conservation.

The NDP's attempted re-regulation of forest policy experienced problems in its own terms of reference. For example, the NDP felt obliged to reduce stumpage levels in 1997, even though they remained relatively high, as the BC forest economy experienced a severe recession not felt elsewhere in Canada. ENGOs, however, saw this move as a 'climb down' and a sign that industry continued to have undue influence on provincial policy. Indeed, ENGO anger led to support for the Green Party and help towards the annihilation of the NDP in the 2001 election (the NDP was reduced from governing party to two sitting members).

The new liberal government quickly modified aspects of NDP policy but has not systematically reversed environmental commitments. The central coast agreement (Great Bear Rainforest to the ENGOs) has been honoured, certification processes are moving ahead, the Code has been streamlined but is still intact, stumpage remains high, and the Treaty Process is a continuing, albeit problematic endeavour. Moreover, the present government is implementing a further substantial reform of the tenure system by reducing the timber held by major timber licensees by 20 per cent to supply new community forests, timber markets, wood lot owners and aboriginal treaty rights. Economically, both the NDP's attempted re-regulation and the liberal's timber policy reforms are seeking to stimulate a higher value, more diversified environmentally friendly range of industries that is viable despite higher wood fibre costs.

Yet, the extent to which BC's forest economy has restructured towards a value added future remains in question. In 1998, even as they recognised some positive developments, Ernst and Young (1998, A10) still found the industry to be "trapped in the commodity box mix" and if commodities "will remain the backbone of BC's forest economy … A significant shift towards higher value growth products is also essential" (Ernst and Young 1998, A81). For Wilson (1998) and Cashore *et al.* (2001), this failure to restructure properly and meet environmental obligations is rooted in corporate dominance of timber tenures (Ernst and Young 1998) and the ability to frustrate reform while ENGOs remain of marginal influence. Such views, however, underestimate the sometimes desperate and often controversial attempt corporations have made to restructure themselves over the past two decades (Hayter 2000). These views also underestimate the role of ENGOs as power brokers in BC's forest economy. Indeed, I suggest that ENGO tactics themselves have negatively influenced corporate restructuring processes in unintended ways.

An Unappreciated Implication of the War in the Woods

ENGO belligerence has consistently been directed towards large-scale industrial forestry and commodity production. From time to time ENGOs have indicated support for a restructuring of industry away from its commodity orientation towards higher value activities, especially if performed by small and medium size enterprises (SMEs) (M'Gonigle and Parfitt 1994). Higher value activities are seen as more labour intensive but less fibre intensive while SMEs are considered to have greater flexibility in using smaller, more varied wood supplies to serve smaller more varied markets. In this regard, there is at least an implicit assumption that industry will be able to produce product values that can compensate for the higher costs of increased stumpage, environmental regulations and less wood fibre availability. This assumption may be regarded as 'heroic' and not based on evidence.

The potential of BC's forest industry to respond to escalating costs is not widely understood. Similarly, the potential supply of entrepreneurs capable of adding significant value to forest products is unknown. New regulations may also place additional burdens on SMEs because of scale effects. Moreover, in the face of higher costs and more bureaucracy firms have the option of deepening their search for cost reduction through large scale technology rather than by adding value to outputs by smaller scale more flexible technology. Further, the war in the woods has imposed considerable uncertainty on firms, large and small, as to the nature of future costs. ENGOs, the CFCLI and Aboriginal Peoples have constantly escalated their demands and forest policy has been in a state of flux for decades. Such uncertainty constitutes significant disincentives to invest in long term research and development, innovative behaviour and skill formation.

An important unintended effect of environmentalism on BC's forest economy, related to ENGO belligerence, environmental regulation and bureaucracy, is that the restructuring of BC's forest economy towards higher values and greater diversity may have become less likely. Indeed, the war in the woods has endured for over two decades and belligerence seems very close to the surface. The announcement in 2003 by Canfor, BC's biggest corporation, of a new 'super sawmill', planned as the most efficient in the world, reflects a broadly based trend to reduce costs to stay competitive. Canfor has also invested in several value added mills in the US, as have several other BC firms. Secondary (higher value) manufacturing within BC's forest economy is greater now than in 1980 (Rees and Hayter 1996; Reiffenstein *et al.* 2003) but growth has levelled off. BC's share of the high value Japanese market has declined recently. MacMillan Bloedel, BC's leading innovator, has been acquired and its research and development activities have been closed down. BC's forest economy may be looking back to the future, a direction in which ENGOs may have played an ironic role.

Conclusion

Even if hard to quantify, ENGOs have played a high profile, relentless role in encouraging BC's forest economy towards an environmental agenda. Their campaigns presenting their views and their ecocentrist representation of the environment have achieved remarkable success. BC now has more trees, more parks, environmental certification is expanding rapidly, there are more rigorous controls on forestry, air and water pollution has been greatly reduced, clear cuts are much smaller, and a value-oriented forestry has emerged. Surely such trends should be the basis for self-congratulation among ENGOs? At the same time, the forest economy is highly uncertain and unruly. The forestry industry has been highly volatile, forest communities are in decline, unemployment is increasing, while trends towards value forestry – the use of fewer timber inputs to make high value products – is itself threatened. In this regard, ENGOs might be more self-critical. It may be that ENGO belligerence has itself become part of the problem in the restructuring of BC's forest economy to a more sustainable higher value future.

References

Arnold, R. (1999). *Undue Influence*. Bellevue: Free Enterprise Press.

Cashore, B., Hoberg, G., Howlett, M., Rayner, J. and Wilson, J. (2001). *In Search of Sustainability: British Columbia's forest policy in the 1990s*. Vancouver: UBC Press.

Ernst and Young. (1998). *Technology and the B.C. Forest Products Sector*. Vancouver: Science Council of British Columbia.

Hayter, R. (2000). *Flexible Crossroads: The restructuring of British Columbia's forest economy*. Vancouver: UBC Press.

Hayter, R. (2003). 'The war in the woods': Post-Fordist restructuring, globalization and the contested remapping of British Columbia's forest economy, *Annals of the Association of American Geographers*, 93, 706–729.

Hayter, R., Barnes, T. and Bradshaw, M. (2003). Relocating resource peripheries to the core of economic geography's theorizing: Rationale and agenda, *Area*, 35, 15–23.

Hayter, R. and Soyez, D. (1996). Clearcut Issues: German environmental pressure and the British Columbia forest sector, *Geographische Zeitschrift*, 84, 143–156.

Lee, R. G. (1994). *Broken Trust, Broken Land – Freeing ourselves from the war over the environment*. Wilsonville, Oregon: Book Partners.

M'Gonigle, M. and Parfitt, B. (1994). *Forestopia: A practical guide to the new forest economy*. Madeira Park: Harbour Publishing.

Moore, P. (2000). *Greenspirit: Trees are the answer*. Vancouver: Greenspirit Enterprises.

O'Riordan, T. (1976). *Environmentalism*. London: Pion.

Paehlke, R. (1996). Environmental challenges in democratic practice. In W. M. Lafferty and J. Meadowcroft (Eds.) *Democracy and the Environment*. Cheltenham: Edward Edgar.

Percy, M. B. (1986). *Forest Management and Economic Growth in British Columbia, A study prepared for the Economic Council of Canada*. Ottawa: Supply and Services.

Princen, T. (1994). NGOs: Creating a niche in environmental diplomacy. In T. Princen and M. Finger (Eds.) *Environmental NGOs in World Politics*. London: Routledge.

Read, N. (2004). Half of B.C. needed to save grizzlies, report says, *The Vancouver Sun*, 12 February.

Rees, K. and Hayter, R. (1996). Flexible specialization, uncertainty and the firm: Enterprise strategies and structures in the wood remanufacturing industry of the Vancouver Metropolitan Area, British Columbia, *The Canadian Geographer*, 40, 203–219.

Reiffenstein, T., Hayter, R. and Edgington, D. (2002). Crossing cultures, learning to export: Making houses in British Columbia for export to Japan, *Economic Geography*, 78, 195–220.

Stansbury, W. T. (2000). *Environmental Groups and the International Conflict over the Forests of British Columbia, 1990 to 2000*. Vancouver: SFU-UBC Centre for the Study of Government and Business.

Taylor, R. P. (1996). Democracy and environmental ethics. In W. M. Lafferty and J. Meadowcroft (Eds.) *Democracy and the Environment*. Cheltenham: Edward Edgar.

The Ecologist (2002). Resource wars: Fighting multinationals across the world, *The Ecologist*, 32, 26–27.

Wilson, J. (1998). *Talk and Log: Wilderness politics in British Columbia*. Vancouver: UBC Press.

3. RE-NEGOTIATING SCIENCE IN PROTECTED AREAS: GRIZZLY BEAR CONSERVATION IN THE SOUTHWEST YUKON

DOUGLAS CLARK AND D. SCOTT SLOCOMBE

Introduction

In Canada, as elsewhere, science has long held an important but difficult role in presentations and representations of the environment in environmental management and natural resource conservation. Notable Canadian environmental controversies in which science has been contentious include hazardous waste facility siting, pesticide spraying, wild fisheries management, forestry, food safety, and more recently salmon aquaculture (e.g. Leiss 2000; 2001). The issues raised revolve around the objectivity of government, citizen, and NGO science; appropriate forms and processes for science in applied or mandated science and policy contexts (c.f. Salter *et al.* 1988), the appropriate participants in such processes, the treatment of uncertainties, and the use of the precautionary principle. These issues are prominent in presentations of other regional-scale environmental conflicts in Canada, such as managing wide-ranging wildlife species and preserving 'pristine' ecosystems. Here, we examine questions about the role and influence of science in specific environmental conflicts in Canada's north. There, traditionally natural science-based conservation approaches are being challenged as governance structures and processes change, largely through Aboriginal land claims and self-government, as well as ongoing devolution of federal responsibilities to territorial governments. This evolution in policy processes brings into sharp relief the divergent presentations and representations of the environment by those with different cultural values and worldviews. A substantial literature on this turbulent process of change has also emerged (e.g. Brody 1988; Freeman and Carbyn 1988; Berkes *et al.* 1991; Simmons and Netro 1995; Morgan and Henry 1996; Treseder *et al.* 1999; Nadasdy 2003; Watson *et al.* 2003; Berkes *et al.* in press).

In the southwest Yukon in particular, protected areas and grizzly bears are especially contentious subjects that are interrelated by geography and the history of research and conservation efforts. Applying science to develop viable management regimes for both remains a challenge. This chapter represents our current understanding of these two intersecting management issues in this particular area: it is not

G. Humphrys and M. Williams (eds.), Presenting and Representing Environments, 33-53.

intended as a comprehensive account of either park management or grizzly bear conservation in the Kluane region. This preliminary assessment is based on initial results from ongoing dissertation research by Clark, plus more than fifteen years of environmental planning research in the southwest Yukon by Slocombe (Slocombe 1992; 1993; 1998; 2002; Danby and Slocombe 2002; Danby *et al.* 2003). Research methodology to date includes semi-structured interviews and focus groups with First Nations members, local stakeholders, government managers, and biologists. It also draws on two years experience by Clark as a park warden in Kluane National Park and Reserve. It is important to note that interview data have not yet been fully analysed or validated: the latter is a particularly important consideration given the cross-cultural context of this work. We conclude by discussing how science is evolving in the Kluane region, and how research can be conducted as a process of collaborative inquiry in conflictual and cross-cultural settings.

Complexity, Policy and Post-Normal Science

Our approach to developing knowledge and policy in contested conservation contexts draws on complex systems (Gunderson *et al.* 1995; Kay *et al.* 1999; Gunderson and Holling 2002), policy science (T. Clark 2002), and post-normal science (Funtowicz and Ravetz 1994).

Complex systems theory has multiple origins and broad applications in many fields (e.g. Capra 2002). Here, we draw most extensively on the natural science-based lineages that emphasise resilience of linked social-ecological systems (Holling 1973; Berkes and Folke 1998; Gunderson and Holling 2002; Berkes *et al.* 2003), and managing for sustainability in self-organising systems (Kay and Schneider 1994; Kay *et al.* 1999). Both lineages emphasise change and explore limitations of positivist, reductionist science applied to complex, real-world problems through a mixture of regional-scale empirical case studies and theoretical inquiry. These lineages are critical of existing natural resource management institutions (e.g. Ludwig *et al.* 1993; Holling and Meffe 1996; Ludwig 2001) and promote explicit recognition of value judgements, systemic uncertainty, more participatory involvement, and adaptive learning mechanisms.

Tim Clark's (2002) application of policy sciences methodology (Lasswell 1970) to natural resource management has produced an interdisciplinary problem analysis heuristic intended for application by practitioners seeking to improve their effectiveness within policy processes. This heuristic is useful for investigating contested conservation situations (e.g. Clark 1999; Clark *et al.* 2000; Clark and Gillesberg 2001) and contains five main analytical dimensions:

- problem orientation – a discursive understanding of the parameters of a given problem and possible options for resolution – which broadly corresponds to the components of a rational-comprehensive approach to policy analysis (Stone 2002);

- social process – the aggregate suite of participants, their standpoints, myths[1] and values, as well as strategies they employ and resultant outcomes;
- decision process – the stages of decision making and in reference to the collective set of decisions made by all participants in each case;
- standpoint – clarification of the investigator's own myths, values, and biases;
- knowledge integration – designing a plan of action based on reflective and comprehensive understanding of the situation.

Post-normal science emerged in the 1990s as a response to application of the traditional certainties of positivist science to complex environmental policy problems (Funtowicz and Ravetz 1994). The term deliberately reflects a difference between this approach and Kuhn's (1970) 'normal science' of solving tractable problems ('puzzles') within an accepted scientific paradigm. The justification of post-normal science holds that in situations of increasing uncertainty and/or stakes, normal science is increasingly ineffective, and its usual extension – professional opinion or consultancy – also has limited utility. Funtowicz and Ravetz (1994) argue that, in such situations, the best and most ethical way to reduce uncertainty is through participatory practice; involving an extended peer community of people affected by the outcomes of science to judge its relevance and veracity. From this perspective, co-management of natural resources is an inherently post-normal problem.

Major principles of each of these theories are beginning to be incorporated into the design of current resource management efforts in the Yukon and in the Canadian national park system[2]. Examples include participatory scenario investigations of caribou hunting and management strategies in the north-central Yukon (Kofinas *et al.* 2002a; Smith and Cooley 2003), and understanding stakeholder values about grizzly bears in and around Banff National Park (M. Gibeau, personal communication). Integrating the main implications for practitioners of those three bodies of theory suggests a possible framework for assessing conservation efforts. Such a framework would consider the following points:

- A requirement to understand, respect, and work with other participants' values and perspectives within a negotiated set of social and decision processes (T. Clark 2002), rather than top-down imposition of governmental policy. For detailed examinations of First Nations worldviews in the southwest Yukon – which are particularly important here – see Cruikshank (1998), McClellan (2001) and Nadasdy (2003).
- Such negotiated co-management implies both an extended peer community (Funtowicz and Ravetz 1994) and a shared problem-orientation (Weiss 1989; T. Clark 2002): or at least a process to arrive at one.
- Knowledge integration is often an explicit goal in co-management; especially integrating traditional ecological knowledge (see Berkes 1999) and science (e.g. Freeman and Carbyn 1988; Cruikshank 1998; Nadasdy 2003). Integration is also seen more generally as a way to cope with competing values and hopefully reduce the usually high uncertainty about the system at hand

(Funtowicz and Ravetz 1994) and as a basis for more robust policy options (T. Clark 2002).

- Spatial and temporal cross-scale connections (Gunderson and Holling 2002; Berkes *et al.* 2003) are important features of multi-jurisdictional, ecosystem-based efforts in this region (Danby and Slocombe 2002).
- The challenge of institutional adaptation and social learning (broadly defined), which are key elements in creating more resilient social-ecological systems (Gunderson *et al.* 1995; Berkes and Folke 1998).

Grizzly Bears, Conservation and the Kluane Region, Yukon

When the Alaska Highway was built during the Second World War, concern about over-hunting in the southwest Yukon by the influx of American military personnel led to the creation of the Kluane Game Sanctuary in 1942. This sanctuary was always intended to become eventually a national park, to draw tourism and assist the economic development of the region (McCandless 1985; Lotenberg 1998). In 1976, much of the sanctuary became Kluane National Park and Reserve (Theberge 1980) (Figure 3.1). From the start, park establishment proceeded without local consultation,

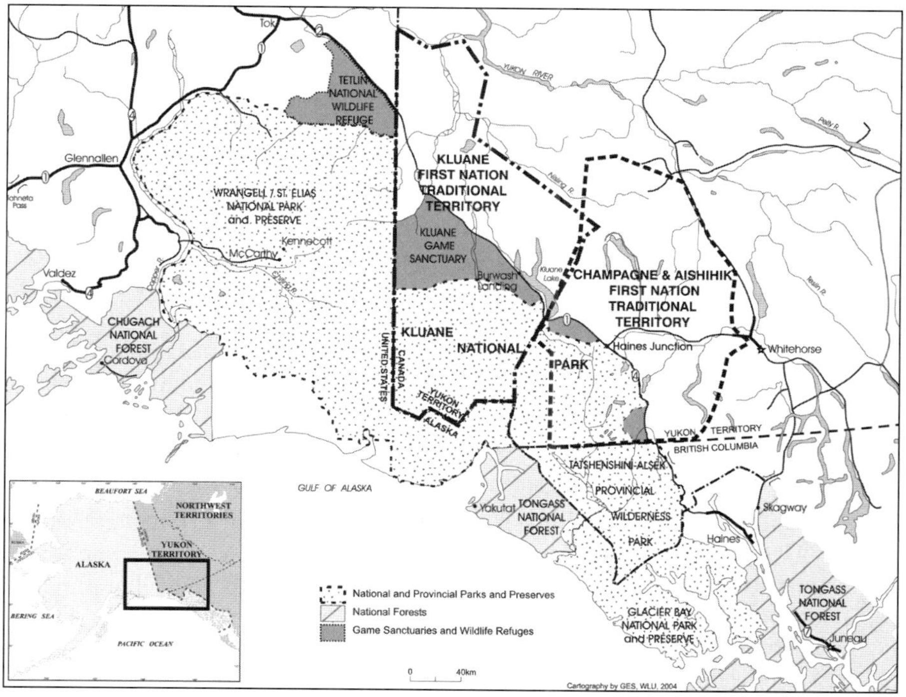

Figure 3.1 *Kluane National Park and Reserve.*

and with considerable subsistence and economic impacts on local Aboriginal and non-Aboriginal people. The memory and legacy of these remain strong in the region, despite economic benefits from park visitors and the region's growing tourism sector. Kluane measures 22,015 square kilometres, contains the highest mountains in Canada, and is located within the boreal cordillera ecozone (Wiken 1986). Only 18 per cent of the park is vegetated, and only 7 per cent is forested: the remainder is largely high-altitude ice and rock (Parks Canada Agency 2004a).

In 1973, Aboriginal People in the Yukon began negotiating land claims with the federal and territorial governments. This resulted in the signing of the Yukon-wide Umbrella Final Agreement in 1993, and, more locally, the Champagne and Aishihik First Nations (CAFN) Final Agreement in 1993. Regionally, the Alsek Renewable Resource Council (ARRC) was created as the primary instrument for renewable resource management in the CAFN traditional territory (Figure 3.1). The Council makes recommendations about management of fish, wildlife and their habitats to the Territorial Government, CAFN Government and the Yukon Fish and Wildlife Management Board (YFWMB). Similarly, the Kluane Park Management Board (KPMB) was created to make recommendations to the park superintendent, and ultimately to the responsible federal minister. The neighbouring Kluane First Nation ratified its Final Agreement in 2003 and the White River First Nation claim is still being negotiated. The signing of those claims will result in new resource councils for those traditional territories and the addition of members to the KPMB from those First Nations. Aboriginal co-management of Kluane National Park is contentious, not least because of the fifty-year history of exclusion of First Nations from the park area after their occupying it for millennia (Lotenberg 1998; Weitzner 2000; D. Clark 2002).

Protecting regionally significant wildlife populations is one of the major goals of the park (Parks Canada Agency 2004a), and grizzly bears have always received considerable management attention in Kluane. One of the first in-depth studies of grizzly bears in Canada took place there in the 1960s, prior to national park establishment (Pearson 1975). Park managers initially considered grizzlies primarily in the context of their conflicts with park visitors (Leonard *et al.* 1990), and managing bear–human conflicts remains a high priority both inside and outside the park (e.g. Wellwood and MacHutcheon 1999). With a growing institutional emphasis on science (Parks Canada 1994), grizzlies came to be seen as an indicator of the park's wilderness values and ecological integrity and a major study on them ran from 1992–1997, with some monitoring activities still continuing (Wielgus *et al.* 1992; McCann 1998; 2001). Outside the park, the territorial government manages grizzly bears as a game animal, with harvest by non-residents regulated through a system of points allocated to outfitters (Smith 1990). McCann (2001) estimated that approximately 225 bears used the park: significantly, 43 per cent of the bears monitored spent time outside the park, and the population appeared to be declining at 3 per cent per year, though there was a wide confidence interval associated with that point estimate (McCann 1998). It is worth noting here that grizzly bears are difficult animals to census (Boyce *et al.* 2001). Not only is there substantial uncertainty about the estimates of regional grizzly bear population size and growth rate, but also those estimates are not of a closed population with clear boundaries, so there is no biological rationale for describing a 'park population' of grizzlies in Kluane.

Following the study of the park, in 2000 the KPMB and the ARRC initiated a regional grizzly bear management planning process intended to provide consistent management among jurisdictions, at an ecosystem scale. The case for such a precautionary, science-based, multi-jurisdictional approach to grizzly bear conservation has been well established by biologists and is widely accepted by management agencies (Herrero 1994; Craighead *et al.* 1995; Mattson *et al.* 1996; Herrero *et al.* 2001; Raufflet *et al.* 2003). Nevertheless, this particular planning process was discontinued within months and has not resumed since.

Discussion

Several key issues and lessons come out of our initial assessment of grizzly bear conservation in the Kluane region. First, there is a considerable conceptual challenge to understanding and acting on each others' values, myths, and perspectives. For example, Aboriginal People and Euro-Canadian people generally have strikingly different concepts of bear-human relationships (Hallowell 1926; Rockwell 1991), which conflict over the appropriate methods and goals of bear research and management (Loon and Georgette 1989; VanDaele *et al.* 2001). The word 'respect' almost inevitably arises in discussions in northern Canada about how to interact with grizzly bears, but as Nadasdy (2003) noted, a First Nations definition of respect is quite different from a non-Aboriginal one, and is rooted in a worldview based on kinship and reciprocity with animals. Even today, these elements of a traditional notion of respect for bears appear to be widespread and strongly held by CAFN members.

In contrast, by the early 1990s Parks Canada had adopted a set of policies rooted in ecological science and conservation biology (Parks Canada 1994). The Kluane grizzly bear study was highly regarded within the Agency and academia because it was designed and led by academic scientists and was guided by a peer review committee of eminent North American bear biologists (Wielgus *et al.* 1992). Their belief was that good science would provide park managers with the necessary information to make good decisions about bear conservation in Kluane.

The differences between these Aboriginal and scientific paradigms led to strong negative reactions to the park's study. Much local concern focused on 'disrespectful' research methods, which were orthodox wildlife study techniques such as chemical immobilisation and radio-collaring, consistent with federal animal care standards (Wielgus *et al.* 1992). It is interesting to note that though Pearson's (1975) earlier study – which is still mentioned in the Kluane region – utilised much the same methods, his work is much less criticised by local residents than the park's is. This outcome most likely results simply from the different social contexts of the two studies: Pearson's work was not affiliated with the park, and community involvement was not demanded of scientists during the time of his study (Yukon Native Brotherhood 1973).

A further underlying difference between these myths may be the scientific tendency to average differences and seek universally generalisable results (Forsyth 2003), versus local inhabitants' awareness of subtle environmental differences within a specific region. This dichotomy may explain why the study results came to been seen by local people as only applying inside the park, but as broadly applicable outside park boundaries by park staff. For example, there are abundant local observations of significant feeding on spawning salmon (*Oncorynchus* sp.) by grizzlies in areas bordering the park yet stable isotope studies including the park study's sampled bears showed no salmon use (Hilderbrand *et al.* 1999). Those data were collected largely in a portion of the park where there are no salmon runs, but more importantly were presented without any spatial description of the sampling: simply the sweeping conclusion that grizzlies in Kluane do not eat salmon. That inference is a generalisation that few people with local experience would accept.

Negative reactions towards the park's bear study were not limited to First Nations people and reduced its legitimacy in the local communities, making it an unlikely foundation for policy decisions by the relatively new community-based institutions. This diminished legitimacy became important because, while the results of the study suggested that the grizzly population was in decline, the local perception was that the results were not only obtained with questionable methods but also were very much at odds with their own perceptions of grizzlies being abundant and simply not in any imminent danger. A common explanation by local people was that if they believed that grizzlies were threatened in the region they would certainly support conservation measures, but since they were not convinced that was the case they saw no need for such action: especially when there were higher priorities, such as restoring important subsistence species including moose (*Alces alces*) and salmon, and coping with forest fire risk to communities, and limited resources to address them.

Funtowicz and Ravetz (1994) describe just such a legitimation crisis from applying the standards of normal science in a post-normal context: "if governments try to base their appeals for sacrifice on the traditional certainties of applied science, or on the authority of professional consultancy, this will surely fail to carry conviction" (p. 1884). Faced with a developing legitimation crisis, park managers and biologists made genuine efforts to communicate their perspectives to other actors, but most still seemed reluctant to question publicly their own individual and collective standpoints. Their perspectives hardened into a position in defence of their research, and the institutional counter-response is summarised in McCann (2001): "the frequent appearance of sub-adult grizzly bears along highways and near communities strongly influences perceptions as to the grizzly bear densities throughout the entire ecosystem" (p. 2).

This representation attempts to explain local viewpoints contrary to the study results, but it is speculative and likely insufficient, since the elements and processes involved in the formation of local perceptions about grizzly bear numbers have not so far been identified. In essence, neither camp believes that the other has an accurate understanding of the status of the grizzly bear population in the region. No common ground has yet been found between these divergent standpoints, and two competing narratives currently exist in the region: a local perception of abundant bears and governments with protectionist agendas, versus the more pessimistic narrative of near-inevitable decline of exploited grizzly populations (Schwartz 2001) and an ill-informed populace, by many scientists and policymakers.

These observations lead to the second lesson: the related challenge of how integration of science with local knowledge and traditional ecological knowledge can be achieved through existing co-management frameworks. Such highly participatory (and at least nominally pluralistic) social processes are laudable attempts to create extended peer communities, though the supposition that they should produce integrative, or even representative, policies may not be realistic. Instead, these processes may serve as arenas for competing problem definitions, with outcomes determined through actors' strategic utilisation of base values (T. Clark 2002) such as power, respect, enlightenment, and rectitude. The collapse of the Kluane regional bear management planning process in 2000 can be seen as a divisive, contentious displacement of the scientific and governmental problem definition by a local one.

A second apparent example was the intervention of the Yukon Fish and Wildlife Management Board (YFWMB) in the acrimonious predator control debate of the 1980s and 1990s: precisely at the time when the land claims of the First Nations in the Yukon were being settled. The territorial government proposed a very large-scale bear and wolf (*Canis lupus*) control experiment between Whitehorse and Kluane National Park to increase moose populations; based on the findings of Larsen *et al.* (1989) that grizzly bears were major predators of moose calves there. The YFWMB surveyed local First Nations and discovered that their concerns were not with bear predation on moose south of the Alaska Highway, but with past overhunting and presently high wolf populations north of the highway. Though the precise sequence of events is not yet clear, interview data suggest a possible causal link between the YFWMB survey and the territorial government's subsequent shift to controlling wolves in the Aishihik area, north-east of Kluane National Park, to assist the recovery of both moose and caribou (*Rangifer tarandus*) populations there (Hayes *et al.* 2003). Whether or not that was the case, the territorial government apparently was also uneasy about how any direct grizzly bear control efforts would be received by the public and interest groups. Nevertheless, even though this programme remained controversial within the Yukon, in Canada, and internationally, it was conducted with significantly more public input than earlier Yukon predator control efforts (McCandless 1985; Hayes *et al.* 2003).

Rather than a complementary blending or recognition of mutual strengths, these two examples each show a process of competing claims for legitimacy among different knowledge bases, with one winning out over the other; at least in the short-term. The question arising from these outcomes is what is really meant by knowledge integration? Expectations of integration in co-management processes are rarely spelled out in detail. Defining knowledge integration simply in terms of blending pieces of information from different knowledge bases does raise questions of purpose and power relationships in the policy process. In his critique of wildlife co-management in the southwest Yukon, Nadasdy (2003) asserts that knowledge integration generally consists of codifying and distilling traditional ecological knowledge into artifacts, such as de-contextualised quotes and maps, that are then simply incorporated into existing state wildlife management systems, thus continuing to marginalise Aboriginal Peoples' fundamental concerns.

The examples presented here suggest another picture, in which the balance of decision making power is actually shifting to favour wildlife co-management bodies – at times generating outcomes that may be more acceptable to local people than they may be to governmental resource managers who are accustomed to the privileged role of expert. This admittedly preliminary conclusion might be seen as contrary to Nadasdy's (2003) instrumentalist interpretation of existing power relationships in wildlife co-management. While this difference can be superficially explained by our specific subjects of analysis – Nadasdy primarily considered issues about hunting Dall sheep (*Ovis dalli*) and moose outside the park – our diverse and contrasting outcomes reinforce the idea that evolving co-management systems are dynamic, complex, and heterogenous. The two positions are not necessarily incompatible either: the first assesses what happens to knowledge as such, the second focuses on broader, contextual, power and institutional relationships.

From the perspective of post-normal science, perhaps knowledge integration means less of a blending than an open, equitable consideration of not just system uncertainties but also value sets and decision stakes. In other words, the result of knowledge integration is not a hybrid knowledge product, but a redefinition of the policy process itself (Hajer 2003). Such a policy process would likely have less predictable outcomes, and the negotiation of a new problem definition would not be surprising. The quality and utility of such policy outcomes must also be negotiated rather than judged 'objectively'; though this does not imply retreating into relativism and parochialism, as many environmental scientists seem to fear (Soulé and Lease 1995). The emphasis in assessing quality in post-normal science is on capacity: the approach is concerned with the legitimacy and effective adaptation of governance processes over time. Such a perspective departs markedly from the traditional technocratic paradigm of resource management; where experts determine quantitative management objectives that bureaucracies then mechanistically implement, with little attention towards either legitimacy or adaptability (Holling and Meffe 1996; Ludwig 2001). No transition between those qualitatively different approaches should be expected to be rapid or smooth.

A third lesson is the need for multi-scale and trans-boundary approaches to defining problems. Such a cross-scale perspective reduces the chance of being surprised by system drivers that operate at different spatial or temporal scales than the system components being focused on, but have been 'externalised' by reductionist research or management approaches (Gunderson *et al.* 1995; Gunderson and Holling 2002; Berkes *et al.* 2003). The importance of interagency co-ordination is well-recognised in ecosystem-based management processes in the Kluane region and the neighbouring St. Elias region of Alaska, and examples of such co-ordination exist at various organisational levels and geographic scales (Danby and Slocombe 2002). Spatial boundaries seem to be a major part of the problem here: grizzlies regularly move in and out of the park area (Pearson 1975; McCann 1998) so the region probably supports a single and likely very widely ranging biological population that is shared among jurisdictions with quite different goals. For example, bears involved in conflicts with humans inside Kluane National Park are sometimes intensively managed with the hope of modifying their behaviour, whereas outside the park (where fewer trained personnel are responsible for a much larger area) the general response is simply to shoot the offending bear. Approximately five such control kills per year are reported and this level is a concern for population sustainability (McCann 2001; Parks Canada Agency 2004a). Conversely, in an attempt to enhance the moose population, the territorial government liberalised grizzly bear hunting opportunities in a large area bordering the park after Larsen *et al.*'s (1989) study (National Research Council 1997). That effort contrasts with the regime in the park – literally right across the road – where hunting of grizzlies is prohibited. Not surprisingly, at times these diverging priorities and approaches have placed the agencies at odds, at least at the senior management level, and overshadowed the otherwise amicable field-level interagency relationships.

Despite those differences, there have been a number of notable successes in bear management at the local scale that cumulatively have probably had region-wide effects. Most of these initiatives were aimed at controlling access of bears to human garbage: a major cause of bear-human conflicts and human injuries by bears throughout North America (Herrero 1970; 1985). In the southwest Yukon, these loosely-co-ordinated (but mutually endorsed) efforts involved territorial conservation officers closing many small landfills along the Alaska Highway, municipal installation of an electric fence around the Haines Junction landfill, and bear-resistant garbage containers installed in all jurisdictions throughout the region – including a custom-built garbage trailer enclosed by an electric fence at Klukshu, a traditional CAFN salmon fishing village. Effective techniques to assess and mitigate the risk of grizzly bear interactions with river rafters were pioneered in Kluane and subsequently implemented in other neighbouring protected areas (Wellwood and MacHutcheon 1999; MacHutcheon and Wellwood 2002). This type of outcome is similar to other regional conservation efforts, such as the Park-CAFN moose harvesting agreement in Kluane National Park and the Greater Kluane Land Use Plan, which are informally implemented at the field level even though they have never been officially approved by senior levels of government. Danby and Slocombe (2002) observe that most interagency co-operation in the Kluane region is similarly informal, and that complex, formal arrangements are rare. While all actors recognise the value of transboundary co-operation and have experience with it in this same region, in this case they have only been able to achieve it at a very grassroots level and where clearly defined interests are compatible. Significantly, these results are consistent with Primm's (1996) case for pursuing small, concrete actions – as opposed to grand plans – towards grizzly conservation in the US Rocky Mountains.

The fourth lesson is that institutions must be able to recognise and adapt to change and surprise in linked social-ecological systems. Kluane National Park seems to face particular challenges as it changes its approach to governance. While staff continuity is often – and rightly – regarded as a strength in organisations, it can also reinforce *status quo* policies and procedures, making change difficult. In Kluane, turnover of field staff has been extremely low, especially in contrast to relatively rapid rotation in senior management positions. Recent structural reorganisations have also concentrated those managers in Whitehorse, 150 kilometres away. These temporal and spatial differences highlight some of the distinctive attributes of the conservation-oriented field culture and the more 'business-like' management culture that co-exist within the Parks Canada Agency, and apparently in national park agencies elsewhere (Wright 1992).

These two distinct occupational cultures generally exhibit different approaches and priorities; likely related to the broad, dynamic range of responsibilities that confront managers, versus the generally more focused and technically-oriented scope of field jobs (Parks Canada Agency 2000). In Kluane, many field staff and managers welcome the park's new emphasis on external co-management partners, but some field personnel feel their own achievements may no longer be valued, and clearly fear to question new directions in case their views are dismissed (D. Clark 2002). Mutual trust between field personnel and management has declined; diminishing morale, making communication difficult, and discouraging innovation and risk-taking at the field level. Similarly, and for a number of possible reasons, some senior managers may have been reluctant to reduce their autonomy by fully empowering co-management partners.

Taken together, these organisational attributes have not only reinforced the divide in Kluane between field and management staff, but also have made the formation of interpersonal relationships with co-management partners more difficult (Weitzner 2000). All of these factors – plus ongoing agency-wide reorganisations and budget cuts (Dearden and Dempsey 2004), and the need to follow national policies and the Canada National Parks Act and regulations – have so far made it extremely difficult for Kluane National Park to adapt fully to the post-land claim environment.

Within this context, the grizzly bear research and management programme in Kluane illustrates the problems of institutional adaptation to changing circumstances. The study was conceived and implemented in the early 1990s, prior to land claim settlement and following the then current organisational emphasis on ecosystem management and better science (Parks Canada 1992; 1994): within Parks Canada this study was 'leading edge' when it was designed. However, once land claims were settled and co-management bodies formed, other actors' expectations of the park were suddenly not just for consultation, but rather for direct involvement in all stages of decision making. The park's well-intentioned attempts to involve First Nations and community members (even inviting First Nations on to the peer review committee) were generally considered too little too late, and so were not viewed as terribly meaningful. Adaptation was further hindered by the unwieldy financial and project management mechanisms which park staff had to use in order to support a large, multi-year research project within an agency that usually budgets on an annual basis. Overall, the project simply had too much momentum to change course rapidly.

A mid-1990s attempt at outreach had unintended effects when a television documentary on the project offended a number of CAFN elders, who felt that the research methods showed a lack of respect for the bears. Park staff had utilised a First-Nation owned film company for the documentary and apparently did not want some of the handling procedures filmed, but felt pressured into allowing it. In 2003 park interpreters developed an outreach programme for local schoolchildren based on playing the role of a radio-collared grizzly bear in the park, which was also poorly received by CAFN members. None of these individual *faux pas* were milestones, but collectively they reinforced a pre-existing local lack of trust in the park's management efforts.

While the park has had some successful adaptive efforts such as engaging CAFN in park management planning (Neufeld 2001), the example of regional grizzly bear management illustrates just how challenging the project of organisational change in a national park actually is. The difficulties of changing a large organisation have implications for resilience of the regional social-ecological system, since the park is a key institutional actor with a large jurisdiction and significant resources, as well as a regional anchor for sustainability and development efforts (Danby *et al.* 2003). Holling and Meffe (1996) describe a pathology of renewable resource management, in which command-and-control institutions that attempt to control ecological variability (managing for a steady-state) sow the seeds of emergent surprises that threaten ecological and social sustainability, as well as the institutions themselves. Kluane National Park is clearly managed for an ecological steady state (Parks Canada Agency 2004a, 30–33) and though it is managed much less hierarchically than in the past, at times still exhibits command-and-control tendencies. While a fluctuating grizzly bear population is unlikely to affect regional sustainability, command-and-control management of other aspects of the social-ecological system such as wildfires, insect outbreaks, or human use conceivably could lead to such surprising and undesirable outcomes.

The final lesson is the difficulty of transferring lessons from elsewhere and the concomitant need for place-based social learning. What is being imported, how, and by whom, affects its reception: here, park staff and researchers promoted a conservation biology-based narrative for grizzly bear management which warns of a slippery slope towards extinction (Schwartz 2001). That deterministic narrative was developed in a profoundly different social and ecological context, although some conservationists point out that those places, such as Banff, Jasper, and Yellowstone, were not very different from Kluane not long ago. Nevertheless, that narrative was profoundly rejected in the regional grizzly bear planning process, explicitly because it was from elsewhere and many local people did not believe that it accurately described their current situation. It would be easy to dismiss their attitude as simply a form of exceptionalism, but doing so would miss the point. Rather, it is a perspective that reflects the deeply held desire for self-determination and control over lands and natural resources, which is a consistent theme in the political evolution of Canada's northern territories (Loveday *et al.* 1989).

No imported idea or theory is likely to be accepted uncritically in the Kluane region, and so progress towards sustainability will likely depend on local-scale experimentation and social learning through direct involvement. Learning in natural resource management regimes appears to more often come from experiencing crises, rather than from examining and adopting 'best practices' from elsewhere (Berkes and Folke 1998). How much learning is actually taking place in the grizzly bear management process is not yet clear. Although many individuals are obviously learning a great deal, very little of that appears to translate into learning at a social level: there is effectively no social learning taking place. The degree of residual suspicion and sensitivity that remains among some actors in the Kluane region suggests that there may even have been some perverse learning outcomes: institutionally convenient but not necessarily valid (Ascher 2001). Diduck *et al.* (in press) suggest that the type of social learning required for change is double-loop learning (Argyris and Schön 1978). In double-loop learning, actors experiencing a mismatch between their intentions and the outcomes of their actions go beyond modifying their behaviour to change the outcome (single-loop learning): they question and modify their 'theories-in-use' – their goals, norms, and assumptions (in policy science terms, their standpoints).

While the actors in the Kluane region clearly identified their conflicts over values and goals, they retreated from that conflict rather than questioning or modifying their theories-in-use. For example, in the absence of a regional bear management plan, park personnel adapted by continuing their incremental approach to site-specific bear management projects: a single-loop learning process. Although this is a pragmatic approach it may have run its course, since most opportunities for bears to obtain human food and garbage have now been eliminated. While learning-based conservation approaches may be theoretically desirable for the situation in the southwest Yukon, they are unlikely to be implemented simply because an imported 'best practices' approach failed to take root there. Given the current climate of distrust of the park and sensitivity about grizzly bears, they may not even be feasible and it may take a crisis or a substantial change of actors before such efforts could be undertaken for grizzly bear management in the Kluane region. In the meantime, perhaps other topics would be more fruitful choices for building co-management experience there: for example, moose harvesting or regional forest management.

Conclusions

What appears to have happened to grizzly bear management efforts in the Kluane region was a rejection by local inhabitants of what they perceived to be an imposition by a centralised government agency of both a predetermined problem definition and a set of policies based on research that was not accepted locally. This rejection occurred within post-land claim social and decision processes where local and First Nations people were sufficiently empowered to be able to affect policy outcomes, and within which Kluane National Park does not appear to have sufficiently adapted to be able to work effectively. While nearly all individual actors are aware not only of this new institutional terrain but also of important components of new resource management theories (e.g. the value of a transboundary approach), the challenges of overcoming history, entrenched positions, and resource limitations remain considerable.

Practical obstacles also remain in the way of inter-jurisdictional collaboration (e.g. legal mandates, perceived policy differences, asymmetry of human and financial resources). Perhaps even more significantly, there still seems to be much misunderstanding about different cultural perspectives on human-environment relationships. Actors imposing their own definitions on others' statements – about respect for grizzly bears, for example – might block cross-cultural communication without even being aware of what is happening.

Given the many challenges to scientific research in this situation, what role remains for science in northern Canada's protected areas, let alone for a scientist in such a new environment? National park policies that emphasise science and a legislated mandate to maintain ecological integrity are relatively recent, and science has never been consistently dominant in the management discourse of Kluane National Park or even elsewhere in the national park system (Parks Canada Agency 2000). At the metaphorical park management table, science has been most often employed by proponents of specific conservation objectives, who often feel compelled to defend ecological integrity (and science) in the face of competing 'human' interests, such as visitor use and facility development. In cases where specific research results directly relevant to a clear decision exist, they can be decisive factors in those decision processes. An example of this in Kluane is the set of management strategies developed to minimise conflicts between grizzly bears and rafting tours along the Alsek River (Wellwood and MacHutcheon 2002; Parks Canada Agency 2004a, 22). However, such situations are rare and in most decisions direct scientific information is lacking. Consequently, inferences from studies elsewhere or appeals to general principles are used, but seem much less likely to influence decision processes. In such cases, the power of generalised or theoretical scientific representations of the environment does not appear to be as strong as empirical, 'objective' representations. Because of this constant negotiation among multiple priorities in parks and other protected areas, environmental conservation advocates (both within the Agency and outside it) have always had to contend with other priorities. Therefore, in some ways the actual role of science in decision processes in national parks – as opposed to an idealised one – may not be appreciably changed during a transition to a co-management regime, where different values, priorities, and knowledge sets must also be contended with.

Nonetheless, aspects of the role of science in Kluane National Park and Reserve are changing. The 2004 Park Management Plan recognises that using both scientific and traditional ecological knowledge in ecosystem-based management is a new approach, and opens the door for an adaptive learning process to accomplish that goal (Parks Canada Agency 2004a, 29 and 33). This suggests scope for some very deliberate and open thinking about the relative roles of science and other sources of knowledge within the parks' co-management system. Such discourse is more likely to occur as ongoing formative evaluation than as a formal summative evaluation process, but developing some institutional mechanism for capturing and disseminating the lessons learned within the Kluane co-management system would significantly assist such learning over time: especially as actors (e.g. board or council members, park managers) change. Other parks in the Canadian north are facing similar challenges as they attempt to apply science to decisions in co-management regimes, and outcomes vary. For example, stable isotope and radiotelemetry studies of Arctic charr (*Salvelinus alpinus*) populations in Quttinirpaaq National Park, Nunavut, have led to clear and broadly-supported recommendations for limiting charr fishing in lakes there (Sahanatien *et al.* 1998; Babaluk *et al.* 2001). In contrast, the question of how to manage access into the polar bear (*Ursus maritimus*) maternity denning area in Wapusk National Park, Manitoba, centres on the contested effects of tourism and scientific research on polar bears (Parks Canada 1999; Lunn *et al.* 2002). Without data on the effects of most activities, developing and implementing regulatory controls for the denning area (Parks Canada Agency 2004b) has been a considerable challenge, and has taken place in a co-management regime where negotiation of the roles and contributions of science and traditional ecological knowledge is a constant feature of the discourse (Clark 2001).

Besides augmenting the existing trend towards interdisciplinary science in the Kluane region (Danby *et al.* 2003), another possible outcome of a deliberative learning process might be a broadening of the kinds of scientific investigation undertaken. As research programmes developed in the Kluane region during the latter half of the twentieth century, physical, biological, and anthropological sciences predominated: each initially with strong disciplinary characterisations (Danby *et al.* 2003). A distinct new phase of collaborative, integrative, and less disciplinarily bound studies may be just beginning; investigating topics such as the paleoecology/ethnography/archaeology of melting high-altitude ice patches (Kuzyk *et al.* 1999; Farnell *et al.* 2004; Hare *et al.* 2004) and Kwäday Dän Ts'ínchi – an ancient human body recovered from a glacier (Beattie *et al.* 2000; Dickson *et al.* 2004), climate change impacts on ecosystems (Danby 2003) and on the resilience of communities (A. Ogden, personal communication), and social and cultural factors in wildlife management (our present study). The Ice Patch Research Project in particular is well advanced and shows how community input at the early stages of a project can provide novel perspectives and fundamentally influence the research questions being asked (Hare *et al.* 2004). Collectively, these studies exhibit a consistent theme of integrating the study of humans with that of nature.

In view of this emergent trend, we suggest that the role of a scientist here is now more substantial and challenging than ever, requiring scientists conversant not only with diverse ecological and social science methods, but also able to understand other peoples' values and seek to understand reflexively their own perspectives and roles in a dynamic policy process. In terms of research practice, it is probably more fruitful to describe some of the steps towards building an effective and relevant process of collaborative inquiry in a contested, cross-cultural context than simply to list further potential research questions (though there are certainly many of those). Although it is still too early to speculate on the eventual outcomes of our ongoing research, there are identifiable elements of practice that have enabled us to build understanding of how the environment of the Kluane region is perceived, presented, and managed.

An interdisciplinary researcher who is demonstrably not affiliated with any of the major local institutions has significant advantages in a situation where conflict exists, such as Kluane. Foremost, that person is in the privileged position of being able to listen to many peoples' stories – crucial elements in understanding values, myths, and standpoints – without the duty of advancing any particular institutional agenda that might constrain some study participants. For example, Clark consistently found that community members were much more open about concerns with park management policies when he returned to Haines Junction as a university-based researcher than when he was working there as a uniformed park warden.

A scientist in this context must also be aware that to some extent that he or she is perceived to represent the entire scientific community, which may not always have positive associations in some peoples' minds. Therefore, it is incumbent on researchers to realise that they do not have to defend all past scientific endeavours in a specific place: rather, they should try to respect all research participants consistently, and earn and maintain as much trust as people are willing to accord them as an individual. Crafting individual relationships with local people takes time, as both authors have experienced, but must be an important long-term goal in any research programme in similar contexts.

Multiple visits over several years and participation in community activities can allow researchers to build collaborative relationships with study participants and engage in extended dialogue, leading to better contextual understanding and potentially to novel forms of collaboration (Fox 2002; Kofinas *et al.* 2002b). Recently formed Aboriginal governments and co-management institutions face substantial and competing demands on their resources, and a researcher cannot expect that his or her own priorities will be the same as those of people they seek to work with. Given this, an important ethical element of collaborative research efforts here and elsewhere in northern Canada is that research must be clearly relevant to local people by meeting real needs that they identify. It is in this post-normal science context that the benefits of collaboration emerge. Relevance and felt need promote full engagement of local study participants and collaborators, and provide important incentives for them to offer their observations, knowledge, and practical assistance to the researcher. A reciprocal obligation exists for the researcher to develop and communicate results that collaborators can readily use and that participants and other community members can understand. In our specific situation, a key element in the emergence of present collaborative research efforts with CAFN has been the relevance of Clark's dissertation research results to upcoming wildlife management planning for the CAFN traditional territory. Collaborative research, especially between cultures in the Canadian north, involves the researcher giving up a certain amount of control: something which may not come naturally to scientists and is only just beginning to be understood and supported by national-level research institutions. However, accepting this shift is a necessary element of collaboration: without it, a research programme would simply be an imposition of yet another outside agenda and would achieve minimal results compared to its potential.

The research process is thus transformed from the essentially solitary and competitive model of modern scientific practice to a communicative process of knowledge co-production (Kofinas 2002b). Scientific information generated by such efforts may take on quite a different role in subsequent decision processes, and could also potentially change social processes by, for example, altering the base values and strategies employed by different participants in particular situations. The role of a scientist in such a re-negotiated context is not a familiar one; indeed there is not even a recognised descriptive term for such work. Linking this kind of research with conservation practice in the southwest Yukon or elsewhere will undoubtedly require experimentation, learning, and acceptance of multiple presentations and representations of the environment: all of which would considerably enrich processes of double-loop social learning that appear necessary to make progress towards sustainability in contested conservation contexts.

Notes

1. T. Clark (2002) defines 'myth' in the policy science context not as a fictional story, but as a shared belief system, analogous to terms such as worldview, frame of reference, or paradigm.
2. However, even such partial implementation does not imply significant acceptance of these theories in their entirety (Clark and Doubleday 2003).

References

Argyris, C. and Schön, D. A. (1978). *Organizational Learning: A theory of action perspective.* Reading: Addison-Wesley.

Ascher, W. (2001). Coping with complexity and organizational interests in natural resource management, *Ecosystems,* 4, 742–757.

Babaluk, J. A., Wissink, H. R., Troke, B. G., Clark, D. A. and Johnston, J. D. (2001). Summer movements of radio-tagged Arctic charr (*Salvelinus alpinus*) in Lake Hazen, Nunavut, Canada, *Arctic,* 54, 418–424.

Beattie, O., Apland, B., Blake, E. W., Cosgrove, J. A., Gaunt, S., Greer, S., Mackie, A. P., Mackie, K. E., Straathof, D., Thorp, V. and Troffe, P. M. (2000). The Kwäday Dän Ts'ínchi discovery from a glacier in British Columbia, *Canadian Journal of Archaeology,* 24, 129–147.

Berkes, F. (1999). *Sacred Ecology: Traditional ecological knowledge and resource management.* Cambridge: Cambridge University Press.

Berkes, F. and Folke, C. (Eds.) (1998). *Linking Social and Ecological Systems: Management practices and social mechanisms for building resilience.* Cambridge: Cambridge University Press.

Berkes, F., George, P., and Preston, R. J. (1991). Co-management: the evolution in theory and practice of the joint administration of living resources, *Alternatives,* 18, 12–18.

Berkes, F., Colding, J. F. and Folke, C. (Eds.) (2003). *Navigating Nature's Dynamics: Building resilience for complexity and change.* Cambridge: Cambridge University Press.

Berkes, F., Fast, H., Manseau, M. and Diduck, A. (Eds.) (In press). *Breaking Ice: Renewable resource and ocean management in the Canadian North.* Calgary: University of Calgary/Arctic Institute of North America.

Boyce, M., Blanchard, B. M., Knight, R. R. and Servheen, C. (2001). *Population Viability for Grizzly Bears: A critical review.* International Association of Bear Research and Management Monograph Series No. 4. Knoxville: University of Tennessee.

Brody, H. (1988). *Maps and Dreams* (2nd Edn.). Toronto: Douglas and McIntyre.

Capra, F. (2002). *The Hidden Connections: Integrating the biological, cognitive, and social dimensions of life into a science of sustainability.* New York: Doubleday.

Clark, D. (2001). Science and management in Wapusk National Park, *Research Links,* 9, 2.

Clark, D. (2002). Land claim settlement and ecosystem management in Kluane National Park of Canada: A case study. In *Parks Canada, An Orientation Program On Ecological Integrity – A Call to Action.* Ottawa: Parks Canada.

Clark, D., and Doubleday, N. (2003). *Resilience Theory in Ocean Management: Finding new bridges, avoiding barrier reefs.* Workshop Backgrounder, Ocean Management Research Network Annual Conference, 13–15 November 2003, Ottawa, ON. www.maritimeawards.ca/OMRN/clark.html (accessed September 2004).

Clark, T. W. (1999). Interdisciplinary problem solving: Next steps in the Greater Yellowstone ecosystem, *Policy Sciences,* 32, 393–414.

Clark, T. W. (2002). *The Policy Process: A practical guide for natural resource professionals.* New Haven: Yale University Press.

Clark, T. W., Mazur, N., Begg, R. and Cork, S. J. (2000). Interdisciplinary guidelines for developing effective koala conservation policy, *Conservation Biology,* 14, 691–701.

Clark, T. W. and Gillesberg, A. (2001). Lessons from wolf restoration in greater Yellowstone. In V. A. Sharp, B. Norton and S. Donnelley (Eds.) *Wolves and Human Communities: Biology, politics, and ethics.* Washington: Island Press.

Craighead, J. J., Sumner, J. S. and Mitchell, J. A. (1995). *The Grizzly Bears of Yellowstone: Their ecology in the Yellowstone ecosystem, 1959–1992.* Washington: Island Press.

Cruikshank, J. (1998). *The Social Life of Stories: Narrative and knowledge in the Yukon Territory.* Vancouver: UBC Press.

Danby, R. (2003). A multi-scale study of treeline dynamics in southwestern Yukon, *Arctic,* 56, 427–429.

Danby, R. K. and Slocombe, D. S. (2002). Protected areas and intergovernmental cooperation in the St. Elias region, *Natural Resources Journal,* 42, 247–282.

Danby, R. K., Hik, D. S., Slocombe, D. S. and Williams, A. (2003). Science and the St. Elias: An evolving framework for sustainability in North America's highest mountains, *The Geographical Journal,* 169, 191–204.

Dearden, P. and Dempsey, J. (2004). Protected areas in Canada: Decade of change, *The Canadian Geographer,* 48, 225–239.

Dickson, G. D., Richards, M. P., Hebda, R. J., Mudie, P. J., Beattie, O., Ramsay, S., Turner, N. J., Leighton, B. J., Webster, J. M., Hobischak, N. R., Anderson, G. S., Troffe, P. M. and Wigen, R. J. (2004). Kwäday Dän Ts'ínchi, the first ancient body of a man from a North American glacier: Reconstructing his last days by intestinal and biomolecular analyses, *Holocene,* 14, 481–486.

Diduck, A., Bankes, N., Clark, D. and Armitage, D. (In press). Unpacking social learning in social-ecological systems in the north. In F. Berkes, H. Fast, M. Manseau and A. Diduck (Eds.) *Breaking Ice: Renewable resource and ocean management in the Canadian North.* Calgary: University of Calgary/Arctic Institute of North America.

Farnell, R., Hare, P. G., Blake, E., Bowyer, V., Schweger, C., Greer, S. and Gotthardt, R. (2004). Multidisciplinary investigations of alpine ice patches in Southwest Yukon, Canada: Palaeoenvironmental and palaeobiological investigations, *Arctic,* 57, 247–259.

Forsyth, T. (2003). *Critical Political Ecology: The politics of environmental science.* London: Routledge.

Fox, S. (2002). These are things that are really happening: Inuit perspectives on the evidence and impacts of climate change in Nunavut. In I. Krupnik and D. Jolly (Eds.) *The Earth is Faster Now: Indigenous observations of Arctic environmental change.* Fairbanks: Arctic Research Consortium of the United States.

Freeman, M. and Carbyn, L. (Eds.) (1988). *Traditional Knowledge and Renewable Resource Management in Northern Regions.* IUCN/Canadian Circumpolar Institute Occasional Paper No. 23, Edmonton: CCI Press.

Funtowicz, S. and Ravetz, J. (1994). Uncertainty, complexity and post-normal science, *Environmental Toxicology and Chemistry,* 13, 1881–1885.

Gunderson, L. and Holling, C. S. (Eds.) (2002). *Panarchy: Understanding transformations in human and natural systems.* Washington: Island Press.

Gunderson, L. H., Holling, C. S. and Light, S. S. (Eds.) (1995). *Barriers and Bridges to the Renewal of Ecosystems and Institutions.* New York: Columbia University Press.

Hajer, M. (2003). Policy without polity? Policy analysis and the institutional void, *Policy Sciences,* 36, 175–195.

Hallowell, I. A. (1926). Bear ceremonialism in the northern hemisphere, *American Anthropologist,* 28, 1–175.

Hare, P. G., Greer, S., Gotthardt, R., Farnell, R., Bowyer, V., Schweger, C. and Strand, D. (2004). Ethnographic and archaeological investigations of alpine ice patches in Southwest Yukon, *Arctic,* 57, 260–272.

Hayes, R. D., Farnell, R., Ward, R. M. P., Carey, J., Dehn, M., Kuzyk, G. W., Baer, A., Gardner, C. L. and O'Donoghue, M. (2003). Experimental reduction of wolves in the Yukon: Ungulate responses and management implications, *Wildlife Monographs,* 152, 1–35.

Herrero, S. (1970). Human injury inflicted by grizzly bears, *Science,* 170(958), 593–598.

Herrero, S. (1985). *Bear Attacks: Their causes and avoidance.* Piscataway: Winchester Press.

Herrero, S. (1994). The Canadian national parks and grizzly bear ecosystems: The need for interagency management, *International Conference on Bear Research and Management*, 9, 7–22.

Herrero, S., Roulet, J. and Gibeau, M. (2001). Banff National Park: Science and policy in grizzly bear management, *Ursus,* 12, 161–168.

Hilderbrand, G. V., Schwartz, C. C. M., Robbins, C. T., Jacoby, M. E., Hanley, T. A., Arthur, S. M. and Servheen, C. (1999). The importance of meat, particularly salmon, to body size, population productivity, and conservation of North American brown bears, *Canadian Journal of Zoology,* 77, 132–138.

Holling, C. S. (1973). Resilience and stability of ecological systems, *Annual Review of Ecology and Systematics,* 4, 1–23.

Holling, C. S. and Meffe, G. K. (1996). Command and control and the pathology of natural resource management, *Conservation Biology,* 10, 328–337.

Kay, J. J. and Schneider, E. (1994). Embracing complexity: The challenge of the ecosystem approach, *Alternatives,* 20, 32–39.

Kay, J. J., Regier, H. A., Boyle, M. and Francis, G. (1999). An ecosystem approach for sustainability: addressing the challenge of complexity, *Futures,* 31, 721–742.

Kofinas, G., Nicolson, C., Berman, M. and McNeil, P. (2002a). *Caribou harvesting strategies and sustainability workshop proceedings. Inuvik, Northwest Territories, April 15–16, 2002.* Fairbanks: NSF Sustainability of Arctic Communities Project (Phase II).

Kofinas, G. with the communities of Aklavik, Arctic Village, Old Crow and Fort MacPherson. (2002b). Community contributions to ecological monitoring: knowledge co-production in the U.S.–Canada Arctic Borderlands. In I. Krupnik and D. Jolly (Eds.) *The Earth is Faster Now: Indigenous observations of Arctic environmental change.* Fairbanks: Arctic Research Consortium of the United States.

Kuhn, T. S. (1970). *The Structure of Scientific Revolutions* (2nd Edn.). Chicago: University of Chicago Press.

Kuzyk, G. W., Russell, D. E., Farnell, R. S., Gotthardt, R., Hare, P. G. and Blake, E. (1999). In pursuit of prehistoric caribou on Thandlät, Southern Yukon, *Arctic,* 52, 214–219.

Larsen, D. G., Gauthier, D. A. and Markel, R. L. (1989). Causes and rate of moose mortality in the southwest Yukon, *Journal of Wildlife Management,* 53, 548–557.

Lasswell, H. (1970). The emerging conception of the policy sciences, *Policy Sciences,* 1, 3–14.

Leiss, W. (2000). Between expertise and bureaucracy: risk management trapped at the science/policy interface. In G. B. Doern and T. Reed (Eds.) *Risky Business: Canada's changing science-based policy and regulatory regime.* Toronto: University of Toronto Press.

Leiss, W. (2001). *In the Chamber of Risks: Understanding risk controversies.* Montreal/Kingston: McGill-Queen's University Press.

Leonard, R. Breneman, R. and Frey, R. (1990). A case history of grizzly bear management in the Slims River area, Kluane National Park Reserve, Yukon, *International Conference on Bear Research and Management,* 8, 113–123.

Loon, H. and Georgette, S. (1989). *Contemporary Brown Bear Use in Northwest Alaska.* Technical Paper No. 163. Kotzebue: Alaska Department of Fish and Game, Division of Subsistence.

Lotenberg, G. (1998). *Recognizing Diversity: An historical context for co-managing wildlife in the Kluane Region, 1890-Present.* Whitehorse: Parks Canada.

Loveday, P., Hodgins, B. W., and Grant, S. D. (1989). Not quite members: Comparative evolution of the territorial norths in Canada and Australia. In B. W. Hodgins, J. J. Eddy, S. D. Grant and J. Struthers (Eds.) *Federalism in Canada and Australia: Historical perspectives 1920–1988*. Peterborough: The Frost Centre for Canadian Heritage and Development Studies, Trent University.

Ludwig, D. (2001). The era of management is over, *Ecosystems*, 4, 758–764.

Ludwig, D., Hilborn, R. and Walters, C. (1993). Uncertainty, resource exploitation and conservation: Lessons from history, *Science*, 260, 17–36.

Lunn, N. J., Atkinson, S., Branigan, M., Calvert, W., Clark, D., Dodge, B., Elliott, C., Nagy, J., Obbard, M., Otto, R., Stirling, I., Taylor, M., Vandal, D. and Wheatley, M. (2002). Polar bear management in Canada 1997–2000. In N. J. Lunn, S. Schliebe and E. Born (Eds.) *Polar Bears: Proceedings of the 13th Working Meeting of the IUCN/SSC Polar Bear Specialist Group, Nuuk, Greenland. Occasional Paper of the IUCN Species Survival Commission No. 26*. Gland: IUCN.

MacHutcheon, A. G. and Wellwood, D. W. (2002). Reducing bear-human conflict through river recreation management, *Ursus*, 13, 357–360.

Mattson, D. G., Herrero, S., Wright, R. G. and C. M. Pease. (1996). Designing and managing protected areas for grizzly bears: How much is enough? In R. G. Wright (Ed.) *National Parks and Protected Areas: Their role in environmental protection*. Cambridge: Blackwell Science.

McCandless, R. G. (1985). *Yukon Wildlife: A social history*. Edmonton: University of Alberta Press.

McCann, R. (1998). *Kluane National Park Grizzly Bear Research Project*. Interim Report to Accompany the Project Review, 21 and 22 October, 1998. Vancouver: Center for Applied Conservation Biology, University of British Columbia.

McCann, R. (2001). *Grizzly Bear Management Recommendations for the Greater Kluane Ecosystem and Kluane National Park and Reserve*. Haines Junction: Parks Canada.

McClellan, C. (2001). *My Old People Say: An ethnographic survey of southern Yukon Territory. Part 1. Mercury Series, Canadian Ethnology Service Paper 137*. Hull: Canadian Museum of Civilization.

Morgan, J. P. and Henry, J. D. (1996). Hunting grounds: making co-operative wildlife management work, *Alternatives Journal*, 22, 24–29.

Nadasdy, P. (2003). *Hunters and Bureaucrats: Power, knowledge, and aboriginal-state relations in the southwest Yukon*. Vancouver: UBC Press.

National Research Council, Committee on Management of Wolf and Bear Populations in Alaska. (1997). *Wolves, Bears and Their Prey in Alaska: Biological and social challenges in wildlife management*. Washington: National Academy Press.

Neufeld, D. (2001). Traditional knowledge as a contribution to national park ecological integrity, *Research Links*, 9, 17–19.

Parks Canada. (1992). *Towards Sustainable Ecosystems: A Canadian Parks Service strategy to enhance ecological integrity. Final Report of the Ecosystem Management Task Force*. Ottawa: Parks Canada.

Parks Canada. (1994). *Guiding Principles and Operational Policies*. Ottawa: Parks Canada.

Parks Canada. (1999). *Wapusk National Park Management Planning Program Newsletter No. 2, April 1999*. Ottawa: Parks Canada.

Parks Canada Agency. (2000). *Unimpaired for Future Generations? Conserving Ecological Integrity with Canada's National Parks. Report of the Panel on the Ecological Integrity of Canada's National Parks*. Ottawa: Minister of Public Works and Government Services.

Parks Canada Agency. (2004a). *Kluane National Park and Reserve of Canada Management Plan*. Ottawa: Parks Canada.

Parks Canada Agency. (2004b). *Wapusk National Park of Canada Draft Management Plan, March 2004*. Ottawa: Parks Canada.

Pearson, A. M. (1975). *The Northern Interior Grizzly Bear Ursus arctos L. Canadian Wildlife Service Report Series No. 34*. Ottawa: Canadian Wildlife Service.

Primm, S. (1996). A pragmatic approach to grizzly bear conservation, *Conservation Biology*, 10, 1026–1035.

Raufflet, E., Vredenburg, H. and Miller, P. S. (2003). Uneasy guests: the grizzly bear PHVA in the central Canadian Rockies. In F. R. Westley and P. S. Miller (Eds.) *Experiments in Consilience: Integrating social and scientific responses to save endangered species*. Washington: Island Press.

Rockwell, D. (1991). *Giving Voice to Bear: North American Indian rituals, myths and images of the bear*. Niwot: Roberts Reinhardt Publishing.

Sahanatien, V., Reist, J. and Babaluk, J. (1998). How do we protect Arctic char? Using otoliths to study migration patterns and assess stocks, *Research Links,* 6, 1/6/7/12.

Salter, L., Levy, E., and Leiss, W. (Eds.) (1988). *Mandated Science: Science and scientists in the making of standards*. Dordrecht: Kluwer Academic.

Schwartz, C. C. (2001). The paradigm of grizzly bear restoration in North America. In D. S. Maehr, R. Noss and J. L. Larkin (Eds.) *Large Mammal Restoration.* Washington: Island Press.

Simmons, N. M. and Netro, G. (1995). Yukon land claims and wildlife management: the cutting edge. In V. Geist and I. McTaggart-Cowan (Eds.) *Wildlife Conservation Policy*. Calgary: Detselig Enterprises, Ltd.

Slocombe, D. S. (1992). The Kluane/Wrangell-St. Elias National Parks, Yukon and Alaska seeking sustainability through biosphere reserves, *Mountain Research and Development,* 12, 87–96.

Slocombe, D. S. (1993). Implementing ecosystem-based management, *BioScience,* 43, 612-622.

Slocombe, D. S. (1998). Lessons from experience with ecosystem-based management, *Landscape and Urban Planning,* 40, 31–39.

Slocombe, D. S. (2002). Complexity, science and inter-jurisdictional cooperation in two very large regions. In S. Bondrup-Nielsen, N. W. P. Munro, G. Nelson, J. H. M. Willison, T. B. Heimes and P. Eagles (Eds.) *Managing Protected Areas in a Changing World.* Wolfville/ Waterloo: SAMPA/PRFO.

Smith, B. (1990). Sex weighted point system regulates grizzly bear harvest, *International Conference on Bear Research and Management,* 8, 375–383.

Smith, B. and Cooley, D. (2003). *Through the Eyes of Hunters: How hunters see caribou reacting to hunters, traffic, and snow machines near the Dempster Highway, Yukon.* Whitehorse: Yukon Fish and Wildlife Branch.

Soulé, M. E. and Lease, G. (Eds.) (1995). *Reinventing Nature: Responses to postmodern deconstruction.* Washington: Island Press.

Stone, D. (2002). *Policy Paradox: The art of political decision-making (Revised Edn.).* New York: W. C. Norton and Co.

Theberge, J. (Ed.) (1980). *Kluane: Pinnacle of the Yukon. The unique environment of North America's highest mountain range.* Ottawa: Doubleday Canada.

Treseder, L., Honda-McNeil, J., Berkes, M., Berkes, F., Dragon, J., Notzke, C., Schramm, T. and Hudson, R. J. (1999). *Northern Eden: Community-based wildlife management in Canada. International Institute for Environment and Development Evaluating Eden Series No. 2/ Canadian Circumpolar Institute Occasional Paper No. 46,* Edmonton: CCI Press.

Van Daele, L. J., Morgart, J. R., Hinkes, M. T., Kovach, S. D., Denton, J. W. and Kaycon, R. H. (2001). Grizzlies, eskimos, and biologists: Cross-cultural bear management in southwest Alaska, *Ursus,* 12, 141–152.

Watson, A., Alessa, L., and Glaspell, B. (2003). The relationship between traditional ecological knowledge, evolving cultures, and wilderness protection in the circumpolar north, *Conservation Ecology,* 8, 2. www.consecol.org/vol8/iss1/art2 (accessed September 2003).

Weiss, J. A. (1989). The powers of problem definition: the case of government paperwork, *Policy Sciences,* 22, 97–121.

Weitzner, V. (2000). *Taking the Pulse of Collaborative Management in Canada's National Parks and National Park Reserves: Voices from the field.* Winnipeg: Natural Resources Institute, University of Manitoba.

Wellwood, D. and MacHutcheon, G. (1999). *Risk Assessment of Bear-Human Interaction at Campsites on the Alsek River, Kluane National Park, Yukon.* Haines Junction: Parks Canada.

Wielgus, R., McCann, R., and Bunnell, F. L. (1992). *Study Design for Kluane National Park Reserve Grizzly Bear Research Program.* Winnipeg: Canadian Parks Service.

Wiken, E. (1986). *Terrestrial Ecozones of Canada. Ecological Land Classification Series No. 119.* Ottawa: Environment Canada.

Wright, R. G. (1992). *Wildlife Research and Management in the National Parks.* Urbana: University of Illinois Press.

Yukon Native Brotherhood. (1973). *Together Today for our Children Tomorrow: A statement of grievances and an approach to settlement by the Yukon Indian People.* Brampton: Charters Publishing.

4. THE MOORLANDS OF ENGLAND AND WALES: HISTORIES AND NARRATIVES

IAN G. SIMMONS

Today's Environment

Within Great Britain, the moorlands (Fig 4.1) are a reservoir of wild but not natural terrains in a highly developed and densely populated land: about 9 per cent of England and 40 per cent of Wales can be regarded as upland, mostly moor though with some mountain in both. Especially within England, in a matrix of land devoted to urban-industrial purposes and to a largely industrialised agriculture, they are seen as the last refuges of the truly wild. Moors are developed on a variety of rock types but are mostly a series of broad flat-topped interfluves dissected by steep-sided valleys. Even though they reach only to 1000m above sea level at most, the oceanic climate and the characteristic treelessness ensure that they appear very different from the lowlands and are often valued for such qualities. Today's land cover is dominated by low vegetation in which grasses, dwarf shrubs and sedges are dominant. The proportions of these differ with the incidence of soil type, drainage, grazing intensity and fire frequency; at lower altitudes the absence of fire and domesticated animals may mean that colonisation by pioneer trees such as birch (*Betula* spp) and rowan (*Sorbus aucuparia*) takes place. More widespread, however, are large tracts of post-1918 forest of planted exotics such as the Norway and Sitka spruces (*Picea abies* and *Picea sitchensis*) and Lodgepole pine (*Pinus contorta*). The moorlands are grazed by domesticated sheep and, occasionally, ponies and feral goats; impoundments reflect the gathering of the high rainfall for urban-industrial use, and extractive industry usually contributes large quarries (Atherden 1992; Thompson *et al.* 1995). However, the productive function is dominated by sheep, which are the basis of an economy only held in place by large subsidies; shooting of grouse (*Lagopus scoticus*) also attracts large fees. The environment itself is the basis of the pleasure-based economy that generates much more money than pastoralism.

As an environment, the moorland cores have spawned at least two historical narratives: the scientific-historical story in which seemingly objective observers chronicle the changes in flora, fauna and other features of the physical environment,

G. Humphrys and M. Williams (eds.), Presenting and Representing Environments, 55-67.

along with historians who plot the evolution of the economy and settlement. By contrast, there is a story of the human reaction to the features of the environment which builds a picture informed by the imagination and by emotion as much as it is by science and documentary history. Some major elements in both these accounts (providing as it were a natural history and a cultural history) are first outlined separately before considering the synergies and dissonances between them.

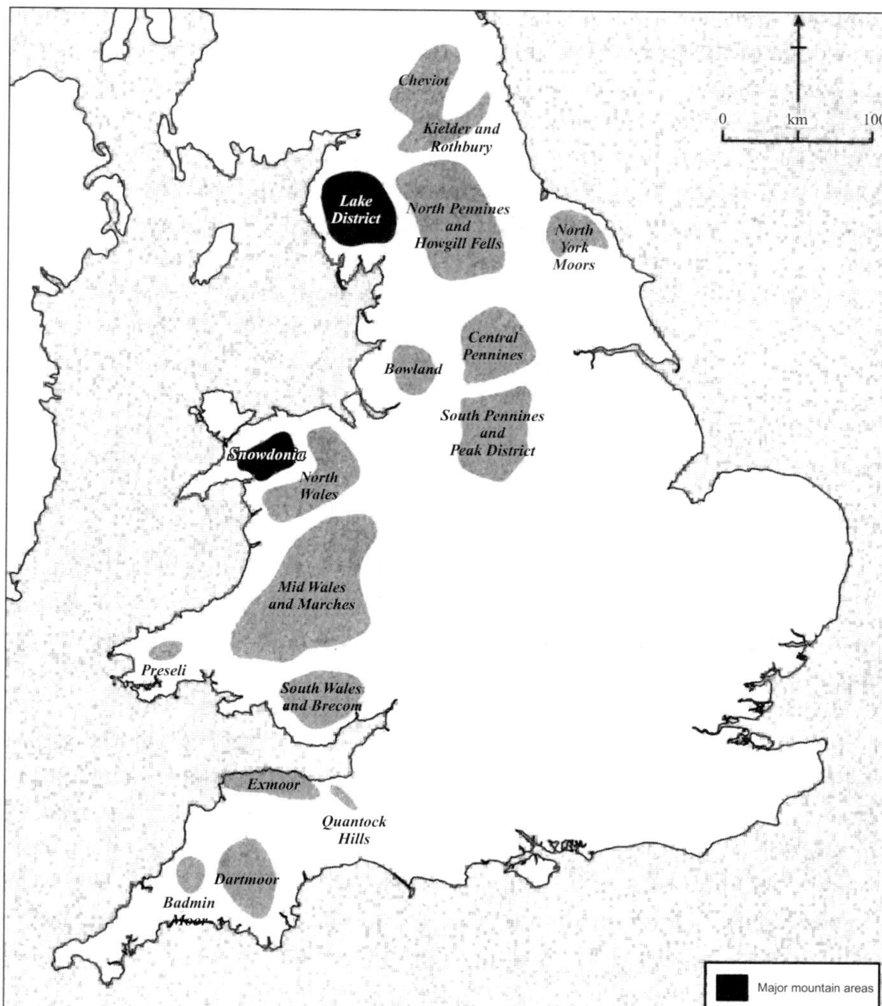

Figure 4.1 *The main areas of moorland dealt with in this chapter. There are some small areas of moor and heath that are only rarely referred to and these are omitted from this map. The two large areas of true mountain are identified but they are not part of the subject matter except in so far as both have peripheries of moorland.*

An Ecological History

The ecology of the moorlands in the Holocene (since about 10,000BC) has always been an interaction between human activity and the features of the physical environment. To simplify these ten thousand years, three phases will be deemed to have been of outstanding importance. These are the mid-Holocene, around 6000BC, the medieval and early modern periods from Domesday Book (AD1086) to AD1600 and the recent past from about AD1800. For each of these, two narratives are possible, though obviously the density of evidence for people's attitudes to the landscape is much higher in recent times; for both the stories some degree of inference and use of proxies is inevitable.

Hunter-gatherers

During the mid-Holocene, these uplands were the habitat of the last foraging cultures (Mesolithic) of Britain, who were superseded after about 4500BC by agriculturalists. The uplands were probably part of a yearly cycle of movement that enabled the people to maximise their access to food resources. Their environment was very different from that of today. The principal dissimilarity was the presence of much more forest: the tree-lines were much higher in altitude, with even the higher summits such as Cross Fell in the Northern Pennines (893m) being covered in a low deciduous woodland. The usual suite of large herbivores of the European post-glacial were found and, presumably, hunted. A major scholarly interest, therefore, has been in trying to determine whether these small groups engaged in any environmental manipulation in order to increase their chances of hunting success.

One likely tool for the ecological modification by humans is fire and the stage is then set for humanised landscapes. Fire does not appear to be responsible for the initial creation of deforested land, rather it was an instrument of keeping the trees at bay once an opening had been created. Hazel (*Corylus avellana*) and grasses have been the major inheritors of space from such processes. Fire is intimately bound up with the disappearance of forest since it perpetuates openings within the woodland; other work has shown that burning took place at the upper edge of the denser woodlands as well. Once the trees have gone, further fire is one of the factors in preventing regeneration. Equally important, the trees acted as water pumps and with their disappearance (probably aided by a climate that was tending to be wetter and cooler) the rate of breakdown of plant matter slowed dramatically, became more acid leading to pan formation in the soils and so peat accumulated. A blanket of peat exists over most moorland above 350m and is usually over 50 centimetres thick and may reach 6 to 8 metres thick in very wet places, even on water-shedding sites (Simmons 1996).

In essence, the Mesolithic folk were the initiators of the openness of today's landscape. They created openings, prevented the tree-line from rising to its climatic potential and started off the growth of extensive blankets of very acid peat whose vegetation and fauna is very limited in species richness compared with the woodland that they replaced.

The Moorlands in an Agricultural Economy

Another critical period was the time from Domesday Book (AD1086) to about AD1600. The moorlands feature in the Domesday Book simply as gaps in the record; some of their regions carry the label 'wasta est' from the devastation of the North carried out by King William I by way of introducing himself to his new subjects. The economy included mineral extraction, and smelting is often charged with the increased loss of woodland in the uplands since feeding the smelting furnaces with charcoal was highly consumptive of trees. But the opposite case needs a hearing since the provision of a sustainable supply of wood was in everybody's interests: coppicing was one means to that end and it is notable that some of the few relict deciduous woodlands in upland England and Wales are single-species stands that could be the remnants of managed woodland of that kind. The moors were the scene of some of the extensive grazing systems which were the foundation of the wool trade. The Cistercian monasteries of the north of England and of Wales were specially prominent, and the development of wool production was aided by the general turn from arable to grass in the fifteenth century (Donkin 1978). The valleys might still produce arable crops and there were still cattle farms (Britnell 2004) but wool production was the ecological nexus in many upland regions. Up on the moors, the typical grassland of acid soils under dense stocking by sheep can become either a moss sward in fifteen years, when it is wet underfoot, or a short tussocky fescue grassland in drier conditions. This may then become dominated by heather in another ten to fifteen years. If grazing pressure is lessened, then the bracken fern (*Pteridium aquilinum*) may invade the pastures at the lower altitudes. In wet conditions, the dominant species may be purple moor-grass (*Molinia caerulea*) which is deciduous and thus provides less nutrition. In order to try to improve conditions for sheep in the earliest part of the season, it became the custom to burn some of these uplands. Where it was wet, this was partly to remove the dead *Molinia* stalks and leaves and allow light through to encourage the early leafing of the acid-tolerant cotton-sedge *Eriophorum* (English Nature 1996). A side-effect of this may well have been faster run-off of water since burning tends to clog the pores of the soils with finely comminuted charcoal.

A major pleasure of the aristocrats in medieval England was hunting. The Normans devoted large areas to the sport and termed them Forests where the hunting rights belonged to the King (Grant 1991). Since the term 'forest' is legal and not ecological, some of the upland moors were Royal Forests: The Peak, Bowland, Dartmoor, Radnor and Pickering are examples; the prize quarry was red deer (*Cervus elephas*). Forests were quite closely managed to make sure that no land uses forbidden by forest law encroached upon the feeding and breeding grounds of the deer. The secondary tasks were to prevent poaching, to see that the habitat was maintained, and to assail predators, especially the wolf (*Canis lupus*). With these successfully achieved then the proper hunt formalities could be observed, with the King taking the stags and lesser nobility undertaking such tasks as the killing of excess hinds.

A synopsis of the agricultural millennia must therefore focus on the role of the economy in keeping the uplands free of trees. In prehistoric times, cultivation extended to higher altitudes than subsequently, but overall the influence of grazing (and notably of sheep) has been dominant in keeping trees down.

Production and Pleasure in the last Hundred Years

As in so many places, the industrial revolution of the eighteenth–nineteenth centuries was immensely pervasive in its influence upon the economy and the ecology of the moorlands of England and Wales. For example, the lower zone of the open land still has many broken-down fences and walls that date the expansion of lower, enclosed, land to the period of high grain prices during the Napoleonic Wars. Many valleys have water impoundments from times when towns sought to improve their supplies for industry and/or their burgeoning industrial populations. Later, after the First World War, large state forests were established on the uplands since the land was cheap. This was also the era of increased public demand for access, especially by walkers. They came into conflict with some of the established uses of the moorlands and none more so than the shooting of the red grouse, and the use of the moors as training grounds for the armed forces, most notably the army.

The sport hunting of mammals still includes red deer, roe deer, rabbits and foxes but is nowhere a particularly dominant activity, especially when compared with grouse shooting. The eastern side of Britain is drier than the west and so the moorlands of the eastern flanks of the Pennines and the North York Moors have been especially good breeding-ground for the red grouse (*Lagopus scoticus*), one of the few birds endemic to Britain. In the nineteenth century, they were shot by walking over the ground with dogs who put up the birds and the hunters shot them from behind. An improvement of grouse density brought about an ecological revolution on these particular moors (see below). It was accompanied by a social-technological revolution, including the advent of the breech-loading shotgun in 1853 and the development of the railways, so that by the 1840s the newly rich businessmen from the industrial areas could reach Yorkshire overnight. To maintain the grouse moors, the moor is fired in 30–50 metre wide strips at 12–15 year intervals, usually in a dry period between late winter and early spring. This creates the conditions for different ages of heather within a smallish territory and so increases the density of birds. Minor measures may include keeping people away during the breeding season, providing grit to aid digestion in the birds' gizzards and killing ground predators like rats. Research, showing that attacks by hen harriers (*Circus cyaneus*) may account for 91 per cent of all grouse chicks killed in their first six weeks, leads to shooting of the predatory bird (Thirgood *et al.* 2000). The main management problem is a parasitic worm infestation that causes a disease called strongylosis in grouse. This appears to result in cyclic variations in grouse populations, with a period of just over four and a half years in England with some variation between drier and wetter moors. In the nineteenth century all this was serviced by a plentiful supply of cheap labour and was the preserve of the old landed classes and the newly rich of the industrial revolution. Today, the transport revolution has gone further. Days on the grouse moors are now marketed in Germany and the Netherlands: charter jet from Dusseldorf, transfer to the moor by sports utility vehicle, a few hours shooting, cold lunch, all for about

GB£4,000 per day. The aim of grouse moor management is now to produce a high density of birds for the shooting season (which begins on 12 August); they are now shot by being driven over a line of guns positioned behind low turf or stone semi-circular structures called butts. Since the high figures of the 1970s, bags of red grouse have been getting lower. There are signs, therefore, that in the long term this ecological relationship is not sustainable and some moor managers are feeding medicated grit to their birds. In the nineteenth century, protection of the grouse meant that there were strenuous efforts to extirpate all its predators: principally foxes, stoats and weasels, but also other species of bird. Small falcons such as the merlin (*Falco columbarius*) are now rare because of such treatment, though attitudes to predators are in general more tolerant now since their role in culling the less healthy members of the population is now realised. But illegal shooting of larger predatory birds such as the hen harrier is still practised, as is the equally illegal digging-out and killing of badgers (*Meles meles*) where their setts adjoin moorland areas (English Nature 1996).

The lesson from the last one hundred years is that of greater intensity of use (in recent times there have been more sheep than ever before) and extension of the range of uses, notably that of the military and of the leisure activities of city-dwellers. Although a less intensively used set of systems than lowland barley-growing or stock finishing, the uplands have nevertheless felt the effects of the increased amount of fossil fuel that an industrial economy is able to deploy.

Moors in the Mind

In addition to this received record of the nature of the moors and their history there is another set of stories: those which are born from the human imagination and thus lodge in all kinds of media a construction (perhaps we might call it 'image' although that word has very strong visual connotations that are a little narrow) of what the moorlands are like. There is then a set of constructions of the moorland environment that derive from the imagination of writers, adepts of visual media, and producers of sound. To be repetitive, these may have two lives: they may stand alone, without reference to the scientific/historical narrative, or they may feed off it and in turn inform its course. In another framework, imaginative constructions are part of the way in which rural environments are formulated as places, which in turn affects the management of their localities. So, while pictures, prose and poetry may not directly affect the way the uplands are grazed or trodden (for example) they contribute to a national consciousness about these environments and help us to consider how they 'ought' to be used.

Pre-literate Attitudes

We know virtually nothing about the non-material life of later Mesolithic times. We can only use the analogies of richer material remains from other Mesolithic cultures, especially those of north Germany and southern Scandinavia, and of near-recent hunter-gatherer societies. All they will allow us to say is that a myth or set of myths probably linked earth, sky, trees, animals and humans in one flow of the mind and the land. The earth and all its materiality were sensate, some of it benevolent, other parts markedly less so. Encompassed in such a cascade would have been the necessary prohibitions to allow the next generations to persist: the Mesolithic folk of upland England and Wales may well have made 'prayers to the Raven'.

Medieval and Early Modern Times

In medieval times, there was some literacy. Such medieval literature as has come down to us, used with due caution, tells us some of the things that people thought, before today's categories like poetry, fiction, and administration become relevant. Even before the Norman Conquest (which produced a flood of bureaucracy), the anonymous author of *Beowulf* was unhappy with the wild places where the monster Grendel's mother lives: "over the sombre moors … along meagre tracks, narrow, forbidding bridle-paths, uncertainways, and beetling crags, past holes of the water-demons" (Wright edition 2000).

We have to be careful about what 'moors' may mean in this context, though it sounds like the right sort of place. (Grendel is described as a *mære morstaÞa*, a notorious moor-stalker.) Anglo-Saxon poetry does not often mention specific environments except the sea, though species may be enumerated. There is in fact no word for 'nature' (meaning the non-human but excluding the supernatural) in Anglo-Saxon: there was apparently no conception of the natural world that excluded the supernatural.

When the Cistercian revival of the Benedictines came to Britain in the wake of the Norman conquest, then one of the favoured areas for monastic foundations were the valleys that ran up into the moors, in manors that were '*wasta*' or where the population could be moved away from the sacred precincts. In 1132, for example, both Fountains (Yorkshire Dales) and Rievaulx (North York Moors) were established. Far from flowing with milk and honey, the early Benedictines in the North York Moors thought of them (as quoted by Menuge 2000) as, "a place of horror and vast solitude … uninhabited for all the centuries back, thick set with thorns, and fit rather to be the lair of wild beasts than the home of human beings".

Alongside their material production (wool, iron, lead, leather, horses, fish) there was the spiritual myth: their life took place in an allegory of the Temptation in the Desert, in an environment that was as much literary convention and monastic imagination as it was physical hardship. So, while the unfree peasant population might see the moors as being yet another element in a hard life, the monasteries were sure of their spiritual rewards for taming the thorns.

There were, to be sure, some secular views. The well-known Welsh poet of the 14th century Dafydd ap Gwilym is known for his nature poetry but he seems to be a creature of the hillside woods rather than the moors above, though *The Mist* is eloquent on the topic of hill-fog, "Like a cassock of the grey-black air, a very sheet without an end, the blanket of lowering rain, a black weft from afar, hiding the world" (adapted from the translation by Jackson 1951).

One remarkable Middle English source comes from the later fourteenth century when *Sir Gawain and the Green Knight* was written and which has detailed descriptions of hunting of deer, wild boar and fox in an upland environment. This literary source details a fox-hunt (almost uniquely for pre-modern times, compared with deer and boar), with Reynard at the last 'reft of his skin of red' a sight still to be seen on many days in the English winter months.

It would be wrong baldly to oppose lay and monastic views of the upland environment: the latter almost certainly influenced the former and in any case those who compiled any written document were likely producing literature rather than social science. We might though infer that the moors were less favoured than the woods even though they yielded meat, milk, wood, hides, peat for fuel, furze, lead, iron and even a little coal. Somehow the negative strand became dominant so that by Early Modern times most of the evaluation of the moors was negative. It was not uncommon for travel writers to decry these sorts of places. Travelling from Carlisle to Newcastle along the line of Hadrian's Wall, John Tracy Atkyns (in Wilton 1980) was not happy:

> For fifteen miles together we travell'd through one of the most barren places in England, which they call the Fells, not a tree, nor even a Stump to be seen, now & then a straggling sheep appears, that with the utmost difficulty keeps itself from starving, perhaps once in a Year an unthinking Crow or Two fly over it … no Modern would ever think of taking up his Habitation in such a Desart …

The Eighteenth Century and After

Not wanting to be "Moderns … in Desarts", many writers began to echo Jonathan Swift's assertion earlier in the eighteenth century that the essential servant of his country and of mankind would be the man who "could make two ears of corn or two blades of grass to grow upon a spot of ground where only one grew before", and there was often a regional identification (the county was seen as an important unit) of the waste as needing improvement and various measures being applied to it. Entirely predictable, then, are some of the remarks in Arthur Young's tour of the north of England, published in 1771, where in Teesdale, the Earl of Darlington is praised for improving moors near Newbiggin, "which used not to yield a farthing an acre rent; but upon inclosing, and then paring, burning, and liming, sowing with turnips, oats and hard corn and laying down with grass seeds, have immediately been advanced to 7s. 6d an acre, at which rent they now remain". The Earl is regarded as a "spirited cultivator". Teesdale, as he went upstream from Barnard Castle, was "most exquisitely picturesque", but he has little time for the upper Tees unless enclosure had taken place and crops of turnips and oats taken; to allow the land to remain 'waste' when it could be "so easily, quickly, and cheaply improved" was a shame (Young 1967).

Nevertheless, radical change was on the way. By 1770, tours of picturesque districts were all the rage and infected even the detached appraisers of the economy. Young's *Tour* of 1771 divides some of its pages into an agricultural and accounting narrative with an extensive asterisked footnote (taking half of ten pages or so) of purely aesthetic judgement on the qualities of the scenery, some of which near

Barnard Castle was "most exquisitely picturesque". Wales, not surprisingly, was a prime candidate for re-assessment of its landscape. Until the 1770s, the rare visitors expected – and hence saw – topographic perils and climatic hazards. Zaring (1977) points out that for most of the eighteenth century Wales featured little in the travel accounts printed in the *Gentleman's Magazine*, whereas Africa and the Americas were staple fare. When lowland England was the norm for beauty, then upland Wales held no attractions: Defoe was happy to leave the barren mountains for areas like the Severn valley in Montgomeryshire, which he considered the most attractive part of that county. This changed in the 1780s and 1790s when Wales featured in most issues of the magazine since readers were now urged to contemplate ideas of an elevating, awful and magnificent kind. Thus, for example, the waterfalls of the Principality provoked a cascade of topographical description, and received an awesome accolade of approval from William Wordsworth.

Changes in attitude in the eighteenth century were most obviously manifest in the way upland areas became subjects for depiction in paintings and engravings. Many of the latter formed the illustrations to books that chronicled travels or were catalogues of antiquities and so writing about these areas became enmeshed in the new artistic consciousness. The gentry were great consumers of these forms of art, and perhaps especially of the picturesque. Since all art is only in a limited sense mimetic and suspension of disbelief is never complete, there always exists the possibility that the assertions of writers and patterns of painters were to some extent ironic: the creators knew that these patterns were fictive and that they were thus alienated from their true ecological linkages. In that sense, they were just versions of the pastoral rather than deeply connected to the land whose story they told: a charge somewhat surprisingly directed at William Wordsworth by some critics. This is to some extent confirmed by analyses that assert the primacy of learning how to 'understand' landscape scenes so as to make them amenable to shared verbalised reaction: the end-point of landscape art (say several commentators) is words. The second half of the eighteenth century saw a flood of re-appraisal of all the hills, albeit led by the appreciation of the English Lake District rather than a moorland area. The relevance is partly simple: the uplands were likely to produce scenes that conformed to the fashionable modes of art or expression, even though pastoral lowlands were a favourite subject. Within the uplands, however, mountains, waterfalls and craggy ridges were much more likely to appeal than the moorlands. Inspection of many of the books and pictures of the period 1750–1820, for instance, reveals relatively few instances of moors as such. Nevertheless, the energy of the Picturesque was capable of penetrating Romanticism to provide a widespread and popular aesthetic appeal, seen by some interpreters as being at heart a form of consumerism but one that has given us what we now call fine art.

Howard's (1991) quantitative exploration of which landscapes were actually painted suggests that moorland was 'the great discovery' of what he labels the 'Heroic Period' of 1870–1910. It was a taste, he thinks, which developed very rapidly in the last twenty-five years of the nineteenth century and was distinct from the previous fondness for mountains. On Dartmoor, for example, a scene described as dreary in 1800 was reviewed in adulatory terms in 1885 and the local artist William Widgery worked his way up the River Lyd as the fashions changed, getting to the moor by about 1860. Dartmoor later became an exemplar of the idea that watercolour was especially good for moorland: books of travel and topography might have photographs of lowland Devon but move into watercolour for the moor (Le Messurier 2002). Between 1950 and 1980, of eighty pictures of Yorkshire landscapes, twenty-seven were of dales, fells and moors, thus replacing the nineteenth century concentration on the Tees and Barnard Castle. One nineteenth century attraction of the moors was their dreariness: there was a moral to the landscape even if there were no people in view, but between 1919–1950 the bleakness was often replaced by the brighter colours of heather and gorse. This might be read as seeing the moorlands echoing the dull round of industrialism in the first instance, but contributing to the pressure for landscape protection in later decades.

In the twentieth century, no better examples of writing about moorlands can be found than the Welsh poet R. S. Thomas (1913–2000) who largely wrote in English but about Wales. Thomas examines the beauty but also the other side: the low productivity, the impoverished life, the afforestation and the emigration, all of which confirm his view that Wales was being eviscerated by the English. The agriculture is typified by: "the lean patch of land/Pinned to the hilltop, and the cloudy acres,/Kept as a sheep-walk", and afforestation means, "A world that has gone sour/With spruce".[1] His biographer, Justin Wintle (1997), however detects a third phase, of greater acceptance of the here-and-now in the collections after about 1980. But these are less frequently land-related than those of the early delight and its accompanying darker side. Another major poet of recent times is Ted Hughes (1930–1998) who is often cosmically distanced in his accounts: "Heather is listening/Past hikers, gunshots, picnickers/For the star-drift of the returning ice",[2] so he is for the long term and when we come to humans the desolation of an emigrating rural population, as well as industrial decline in the textile industries of the valleys, leads to a disintegration of a former unity. Now only the recreational use is worth noticing. Hughes reminds us that anything may eventually be submerged in a return of the ice age: we may be living in an interglacial. Nested inside this long perspective, Thomas's contribution insists that if we harbour a romantic view of these hill-lands as places of freedom and of difference from the city, then we are part of a probably ephemeral set of attitudes. Both, however, recall an essential marginality in both ecology and emotion.

Fiction writers have often been attracted to moorlands, with the nineteenth century once again pushing these places to the foreground. Emily Bronte's *Wuthering Heights* (1847) is a scene-setter, a fact seized upon by film versions; Elizabeth Gaskell's *The Moorland Cottage* (1850) uses the cottage setting to echo the retreat of the characters, from which their emergence across the unimproved land is not always happy; in R. D. Blackmore's *Lorna Doone* (1869) Exmoor is the setting for the lair of the Doones, who were giants and evil-doers in the footsteps of Grendel. An updated version of most of the bleakness comes in Niall Griffith's recent (2001) novel *Sheepshagger* which takes R. S. Thomas's bleakness rather further, via Psalm 58 as an epigraph. Yet the doyen of all moorland fiction must still be Arthur Conan Doyle's

The Hound of the Baskervilles (1912) with our heroes exiled from the warmth of foggy Baker Street into an open carriage in front of which "rose the huge expanse of the moor [with its] grim suggestiveness of the barren waste, the chilly wind and the darkling sky".[3] Another gift to movie-makers and a resurgence of the 'sublime' of the Romantic movement. Bruce Chatwin's book of 1982, *On the Black Hill,* also made a memorable film set in the Black Mountains of Monmouthshire, though without quite the weaving of environment and character that seems to be possible with poetry and fiction.[4] The same appears to be true of the many television drama series and soaps set in upland areas. Cynics might think that cheap locations are a major factor but the notion of the moorland village or town as a *paysage moralis*, or as more *gemeinschaft* than *gesellschaft* lingers on. Moors are also good places for car advertisements since there is a good chance of showing the vehicle in isolation, an appealing image that almost never actually happens.

Fallout?

The ideas thus generated about moorlands rain out upon the rest of the nation. What they contribute is first of all an image of wildness and of dominance by nature. A review of the National Parks of England and Wales in 1991, dealing mostly with uplands (albeit both mountain and moorland), referred to the essence of such places as lying "in the striking quality and remoteness of much of this scenery, the harmony between man and nature that it displays, and the opportunities that it offers for suitable forms of recreation" (Countryside Commission 1991). To some this even merits the (undefined) term 'wilderness' as an extreme contrast to the cities and the 'new prairies' of East Anglian cereal growing. The landscape must look 'natural', for it is a kind of blank canvas that is being sought. Hence, wildness, openness, asymmetry and homogeneity, height, the freedom to wander at will and the apparent absence of human handiwork are all central to the experience. A quote in Harrison's (1991) book on recreation sums this up: "you feel free that you have access to that. And um, that you can actually just go – you can go right into it! It's like going into the picture … you can walk into the moorland. You can actually go in and you can possess it".

Their second endowment is that of image: of a real (or even ideal) community: moorlands are a favourite scene for television series on rural life. Some very serious people, however, have elevated the hill lands to a virtually mystic status. This was especially the case of Wales, where Sir George Stapledon (1882–1960) championed the cause of more rational land use in the uplands. He wanted to see greater diversity of production systems and in particular the replacement of most bracken land with rye grass and clover mixtures. He was an advocate of a holism in which the land was a source of spiritual qualities and the sustenance of the farm distilled this. He was a passionate champion of rural life as it might be, even to the point of wanting to infuse urban dwellers with some good country genes. Moore-Colyer (1999) argues that the survival, "into the 1990s of Welsh upland communities with their distinctive traditions and culture" owes much to the sagacity and vision of Stapledon. The geographer and anthropologist H. J. Fleure (1877–1969) was likewise convinced the Welsh way of life can be seen as rooted in a moral conviction that the rural districts were a necessary challenge to materialism and industrial values (Gruffudd 1994). Nevertheless, Fleure hoped that afforestation, fisheries and hydro-power might transform rural Wales into a thriving set of communities where an older moral order is re-founded on new technology. A kind of echo is heard in today's espousal of ICT (internet communications technology) as a new saviour of remoter places, though we need to ask whether it is the locally-raised young who are keen to adopt e-work, or whether it is the refugees from the towns.

Dissonances

In a period of concern about biodiversity and its ability to adjust to climatic change, the moorlands present a picture of several thousand years of reduction of species richness as peatlands and heavily grazed grasses have replaced deciduous woodlands. Conifer plantations have done little to improve the picture. Efforts are made to extend native woodlands and to improve the breeding success and the distribution of birds like the red kite and the black grouse. Concern over the biota has led to questioning the effects of increased public access to open moor created by an Act of Parliament in 2000 but at the same time moving towards area subsidies for upland farms rather than headage payments. So the ecology is not thought to be in good condition.

Attitudes towards the moors, on the other hand, are biased towards stasis. The core of the system of National Parks and Areas of Outstanding Beauty consists of moorland areas. Here, the presumption is of development control that allows only minor changes, such as extension of existing deciduous woodland. If (as happened during the 2001 Foot-and-Mouth Disease epidemic and again in 2004 in view of new European Union subsidy patterns) there is talk of vast reductions in sheep numbers then the spectre of the encroachment of 'scrub' is raised: few commentators talk of the positive contribution to biodiversity and habitat variation of areas of 'early successional woodland'. The way in which the visual dominates the debate is emphasised by controversies over electricity generating wind-farms. Generally not permitted in designated areas, even being visible from a National Park is deemed by opponents to be unacceptable.

So there is a major dissonance between the narrative of ecological history and its deployment into today's ecosystems, and the valuation of the moors in the public eye and policies. This is not the place (but see Simmons 2003) to rehearse all the details of the arguments about priorities, but since wildness is clearly so much valued then perhaps any trends towards increasing the number of wild species and the variety of their habitats ought to be welcomed. This means that nature is placed before culture in these areas: a reversal of normal precedence but one that may yet be necessary in many spheres of human endeavour. There is quite a hill to climb before the view opens out.

Notes

1. Adapted from the translation by K. J. Jackson 1951: *A Celtic Miscellany*. Penguin Books, Harmondsworth.
2. 'Heather' (1979). In *Remains of Elmet* (photographs by Fay Godwin). London: Faber and Faber.
3. All the classical novels are available in numerous editions. Griffiths' novel was published in London by Jonathan Cape in 2001.
4. Directed by Andrew Grieve, 1988. Connoisseur Video CR096 (VHS format).

References

Atherden, M. A. (1992). *Upland Britain.* Manchester: Manchester University Press.

Britnell, R. (2004). Fields, farms and sun-division in a moorland region, *Agricultural History Review*, 52, 20–37.

Countryside Commission (1991). *Fit for the Future: Report of the National Parks Review Panel.* Cheltenham: CCCP 334 [The Edwards Report].

Donkin, R. A. (1978). *The Cistercians: Studies in the geography of Medieval England and Wales.* Toronto: Pontifical Institute of Medieval Studies.

English Nature (1996). *State of Nature. The upland challenge.* Peterborough: English Nature.

Grant, R. (1991). *The Royal Forests of England.* Stroud: Alan Sutton Publishing.

Gruffudd, P. (1994). Back to the land: Historiography, rurality and the nation in interwar Wales, *Transactions of the Institute of British Geographers, New Series,* 19, 61–77.

Harrison, C. (1991). *Countryside Recreation in a Changing Society.* London: TMS Partnership.

Howard, P. (1991). *Landscape: The artists' vision.* London: Routledge.

Hughes, T. and Godwin, F. (1979). *Remains of Elmet. A Pennine sequence.* London: Faber and Faber.

Jackson, K. J. (1951). *A Celtic Miscellany.* Penguin Books, Harmondsworth

Menuge, N. J. (2000). The foundation myth: some Yorkshire monasteries and the landscape agenda, *Landscapes,* 1, 22–37.

Le Messurier, B. (2002). *Dartmoor Artists.* Tiverton: Halsgrove/Dartmoor National Park.

Moore-Colyer, R. J. (1999). Sir George Stapledon (1882-1960) and the landscape of Britain, *Environment and History,* 5, 221–236.

Simmons, I. G. (1996). *The Environmental Relations of Later Mesolithic Cultures: The creation of moorland landscape in England and Wales.* Edinburgh: Edinburgh University Press.

Simmons, I. G. (2003). *The Moorlands of England and Wales: An environmental history 8000 BC to AD 2000.* Edinburgh: Edinburgh University Press.

Thirgood, S. J., Redpath, S. M., Haydon, D. T., Rothery, P., Newton, I. and Hudson, P. J. (2000). Habitat loss and raptor predation: Disentangling long- and short-term causes of red grouse declines, *Proceedings of the Royal Society of London, Series B,* 267, 651–656.

Thompson, D. B. A., Hester, A. J. and Usher, M. B. (Eds.) (1995). *Heaths and Moorland: Cultural landscapes.* Edinburgh: Scottish Natural Heritage.

Wilton, A. (1980). *Turner and the Sublime.* London: British Museum Publications.

Wintle, J. (1997). *Furious Interiors: Wales, R. S. Thomas and God.* London: Flamingo.

Young, A. (1967). *Six Months Tour through the North of England 1771.* New York: Kelley [a facsimile reproduction].

Wright, D. (Ed.) (2000). *Beowulf.* Harmondsworth: Penguin Books.

Zaring, J. (1977). The romantic face of Wales, *Annals of the Association of American Geographers,* 67, 397–418.

5. EXPLORATION LITERATURE AND THE CANADIAN ENVIRONMENT: FROM WAY-FINDING TO WAYS OF REPRESENTATION AND READING

WAYNE K. D. DAVIES

"The geography of any place results from *how* we see it, as much as *what* may be seen." (J.Wreford Watson 1969, 10; emphasis added)

Introduction

One of the salient features of Canada's distinctiveness is its unique environment, one that has posed immense challenges to its explorers and settlers. Indeed, Atwood (1972; 1995) has argued that the problem of survival and fear of the environment represents a fundamental motif in much of Canadian literature, paralleling the frontier theme in American literature. At first sight this represents a plausible thesis, since most of the country experiences a winter far harsher than any in Western Europe, the origin of its earliest white explorers and settlers. In addition, major barriers such as the Western Cordillera, the Canadian Shield and the frozen Arctic north made travel and settlement arduous, for, with the one exception of the St Lawrence-Great Lakes system it was historically difficult to penetrate into the continental interior. Yet other literary critics (Glickman 1998) have shown that alternative themes, such as the romantic era aesthetics of beauty, the picturesque and the sublime, have also been used to represent Canadian landscapes, whilst Atwood emphasised that survival was a broad theme that could be extended beyond the environmental issue to deal with other features, such as loss of identity. However, all these interpretations of Canadian environments are based *only* upon literary works which are linked to the cultivation and celebration of feelings and emotions about people and the environment and to questions of representation and meaning in which the land is frequently used in novels and poems as a metaphor for other issues such as identity and nationhood (Kreisel 1968; Ricou 1973; New 1997). But complementing these insights from the field of literature has been the work of literary theorists in the past thirty years. They have shown that the production of meaning from any written text is always fraught with difficulty because the

G. Humphrys and M. Williams (eds.), Presenting and Representing Environments, 69-87.

words we use do not 'reflect' the world as assumed by realist views, rather the words 'condition' how we understand and represent it.

These literary descriptions of the Canadian environment are very different to the empirical-scientific approaches that are followed by most geographers and environmental scientists. As Watson (1983) showed in his Presidential Address to the Institute of British Geographers in 1983, they have traditionally ignored, or deliberately downplayed, feelings about the landscape or using it in metaphorical ways in favour of what was assumed to be an objective, rational, empirical agenda about the so-called 'real world'. In many ways this approach is seen as writing down the characteristics of the environment, the various elements and its related processes, initially from observation and later by more sophisticated measurements and experimentation using scientific methods. For many years the approach was assumed to be dispassionate and objective, and focused on unravelling the scientific origin of various environmental characteristics without much concern for the effect of human activity. Recently, in the pursuit of sustainable environment uses, this has been complemented by ecological approaches. These focus on the way that the land is used or influenced by human activity and incorporate values and attitudes to alternative uses as well as the rights of humans and others. But these works also usually ignore literary views, and are still largely evaluated within a scientific framework.

The scientific-ecological and the literary approaches to the description and interpretation of the Canadian environment clearly represent very different traditions. This can be seen in the divergent content found in most books that claim to deal with the Canadian environment in which alternative approaches are absent. Indeed, just as the historic divide between Anglophones and Francophones has been characterised as historically providing 'two solitudes' in Canada, so the vast majority of writings about the Canadian environment can be seen to have its own chasms of silence between these different approaches. But if we are to really understand the Canadian environment and its representation then *all* approaches to the study and portrayal of the Canadian environment need to be considered in the search for more comprehensive understandings, since they have the ability to influence one another in significant ways.

This chapter will illustrate these ideas using new insights from literary theory and especially the critical practice approach (Belsey 2002) that underlie the new cultural geography (Davies and Gilmartin 2002). These show how we need to view our writings and understandings of the environment as 'constructed representations' (Davies 2003), the creation of which are subject to a whole variety of influences. The observational-scientific approach to environmental description has usually assumed some type of 'realist' or 'mimetic' approach that has been succinctly summarised by one eminent literary critic as follows:

> in the ideology of realism or representation, words are felt to link up with their thoughts or objects in essentially right and incontrovertible ways; the word becomes the only proper way of viewing this object or experiencing this thought. (Eagleton 1983, 136)

In the last thirty years literary theory has shown the limitations of such a view, for words and their meanings are essentially elusive. Indeed most now believe that our cognitive processes, writing strategies and reading 'construct' rather than 'reflect' the world we see and our conclusions about it. This is as true for the apparently objective empirical and scientific approaches as those produced in prose works. But these new ideas were slow to spread into geography until the 1990s though from the 1960s there were signs of dissatisfaction with the empirical, realist views, stimulated by the advocates of the importance of perception in our understanding of '*how we see*' things, as indicated by the quotation at the beginning of this chapter. However, these ideas were usually seen as modifications of the realist empirical views and accounts were still focused upon what Turnbull (2000) has called the "product", in this case our knowledge of the environment, and not "how this product was produced". Moreover, these ideas were rarely linked to the developments in literary theory that have questioned the whole basis of our understanding. Since 1990 this situation has changed, with cultural and historical geographers in particular becoming more aware of the utility of the work of literary theorists. Individuals such as Gregory and Duncan (1999, 2) have described their set of essays on travel writing as the production of a "space of knowledge", as much as the "production of a space of power" through the incursion of Europeans into new lands. In addition, there has been more interest upon *how things are written*, as shown by Clifford in anthropology (1986), an approach used by Barnes and Duncan (1992) to address what they called the "graphy" part of the field of geography, compared to the traditional focus of study on the "geo", or world. However the ways that the world is 'seen' or the 'writing' achieved, are not the only influences that need to be considered. As important in our understanding of the environment is the way that people *read and interpret* the writings (Bonnycastle 1991; Davies 2003); the presence of what amounts to a 'readers plural' provides another level in our understanding of how knowledge is constructed.

This chapter is restricted to what Warkentin (1993) described as the first "discourses" about the land of Canada, namely exploration literature. This consists of the written works of the European explorers of America, for the opinions of the Aboriginal Peoples who first discovered and settled the land have largely been lost. For most of the last century the genre was dismissed by literary critics as having no real literary value; what was considered to be literature was based on a canonical tradition that was linked to privileging some works of prose, poetry or art, especially those regarded as the best examples of particular approaches. Moreover, all too frequently these first opinions about the land by explorers have been regarded as realist or mimetic accounts of some new reality and privileged because they were the first records made. By using the new hermeneutic approaches of literary theorists such assumptions about the pristine accuracy of the environmental comments cannot be upheld.

Trends in Exploration Literature

Until the 1960s the study of exploration was largely the domain of geographers and historians or former military men. They focused upon the twin themes of what may be called the 'exploration product' and 'resource use' approaches. The former focused upon the new lands and routes that the explorer found (Warkentin 1964) and the techniques of defining and mapping the precise location of the places and routes. The latter emphasised the description of the resources present in any area that could be of use to the European discoverers and settlers – and, it must be admitted, often exaggerated their importance. There is little doubt that some readers also valued the exploration narratives for the adventure and heroism that were associated with the journeys, as well as the descriptions of exotic and unfamiliar landscapes and peoples. Such issues were considered to be far less important in academic study than the results of the exploratory endeavours. Moreover, there was an unconscious acceptance that these accounts were essentially realist or mimetic accounts of what was seen and found – at least unless it could be shown empirically, from other evidence or subsequent discoveries, that explorers exaggerated, guessed or even lied about their findings.

During the 1960s this situation changed. One main influence came from the type of approach illustrated by the quotation that began this chapter, which argued that we must not assume that we see the objects we study without any preconceptions or values. Professor James Wreford Watson was a geographer and a poet of distinction who had won the Governor General's award in 1956 under his pseudonym James Wreford. He delighted in creating his own topographic verse that was stimulated by the varied landscapes of Canada and long argued for stronger links between geography and literature, observing: "I have never written an article or a book without an appeal to literature" (Watson 1983, 387). He was one of the few geographers to have been an *active* participant in both the traditions of environmental description through his academic geography and his poetry. Of course, Watson was not alone in his interest in the value of using literature in geography (Simpson-Housley and Mallory 1996), or the importance of perception in geography where geographers such as Lowenthal (1961; 1967) had shown that people in different environments had developed very different ways of describing shape, location, orientation, territoriality, colour and categories of experience.

It is these perception studies that begin a *self-reflection on the nature of geographical knowledge about places.* They inevitably cast doubt upon the realist approach that dominated studies of the environment in geography. The incorporation of ideas from perception research showed how the previous opinions, values, prejudices and dreams of explorers, many of which were far from rational, were essential elements of the processes of how explorers viewed the lands they discovered (Allen 1975b). Nevertheless, despite these insights, the focus of exploration studies still remained upon the 'exploration products' of the voyages of discoveries, mainly upon what may be summarised as 'way-finding', the descriptions of the new environments and the construction of maps to pinpoint locations and to show how others could reach the newly discovered lands.

Entry of the Literary Specialists

From the late 1960s experts in literary studies began to look seriously at the writings of explorers, taking a very different approach to previous work in geography, or history. They focused on the way that explorers 'represented' landscapes, for, as MacLulich (1979) observed in the Canadian context, their narratives imposed 'order' on the places visited and sights seen by the very fact of their writings. The result was the development of studies that showed how explorers 'constructed' their narratives, illustrating the myriad of influences that affected the actual writing, or construction of any exploration narrative. The pioneering work on early descriptions of the Canadian environment by explorers by MacLulich (1979) and McLaren (1984; 1985) has been especially important, although almost ignored by geographers until recently. Such studies have come of age in the last twenty years in major books by individuals such as Campbell (1988), Greenblatt (1991) and Pratt (1992) which provide innumerable examples of the application of these new hermeneutic ideas to many different explorers in a world context and focus on two major themes: on interpretation, and the construction of the narratives.

These new literary approaches owe much to the broader movements in cultural studies and literary criticisms having their origins in the work of French writers, such as Derrida and Barthes. Following the work of Levi-Strauss in the 1920s it had been increasingly accepted that the words used to describe the world are really sign systems, not absolute indicators of some thoughts or sights, so that all words are seen to be unstable and have changing meanings. Hence it is difficult to accept the idea of the realist approach in which one word is definitively linked to one thought. Moreover, the new approaches privilege the role of language in our interpretation of the world around us. In the words of another well known literary critic:

> There is no unmediated experience of the world, knowledge is only possible through the categories and laws of the symbolic order ... language is experienced as a nomenclature because its existence precedes our 'understanding' of the world. (Belsey 1980, 46)

The result is that this critical practice approach to textual study shows that: "reality is not reflected by language but produced by it" (Eagleton 1983, 108).

There is still a huge debate about whether language precedes thought (Whorf 1956; Pinker 1997) or whether there are strong feedback connections between the two. But what is crucial about these ideas is that they have destroyed the belief that descriptions of the external world can ever be objective realist accounts, some direct mirror of some external reality. Applied to exploration narratives it can be argued that the writings of the explorers tell us as much about the previous ideas, cultures and representative practices of the explorer writer as the lands being described. In addition, they can be shown to encode European preferences, agendas and interests and unconsciously legitimise European agendas, interests and dominance.

Now, in some ways these new interpretations may be seen as providing further, and more convincing, theoretical evidence for Watson's view that it is 'how' we see the world that is more important than 'what' we see. One of the earliest examples of this approach in exploration literature studies lies in the way that many of the late eighteenth and early nineteenth explorers in Canada seemed to have unconsciously used the picturesque – one of the aesthetics of the Romantic movement in Britain – in their descriptions of the new landscapes they were seeing and, indeed, were searching for the sights that gave them pleasure (Glickman 1998). The various elements that they sought to paint or to describe were summarised by MacLaren (1985b, 44) as follows:

> elevated prospect, animated foreground, meandering river, sunset, coulisses of tree-covered hills that govern the perspective and direct it to a single vanishing point … (together with) … the ruins motif by scattering driftwood.

But another Romantic period aesthetic, the concept of the sublime, was also used by many explorers and settlers in Canada. Many extreme environmental experiences, such as a storm, mountains, waterfalls, or the severity of winter were initially feared for they could lead to destruction. But the Romantics recognised that if the observer could survive the experience another kind of pleasure or rather excitement emerged, one linked to having overcome the frisson of terror and possible death, producing a new appreciation of the environment. One of the most obvious Canadian examples was represented by Niagara Falls, perhaps the ultimate site at which to experience sublimity, given the size of the gorge, the enormous rush of water over the falls, and its noise, which, initially at least, formed such a contrast to the silence of the forests that initially surrounded it. Today the presence of walls along the gorge, the banning of boats above the falls and restrictions on those sailing under the falls, have sanitised the area and reduced the dangers (Shields 1991; Dubinsky 1998), essentially taming the landscape, although the viewpoints still provide an insight into the terror the falls still evoke in some people. But many other environments could not be as easily tamed. Exploration narratives in Canada are full of examples of areas that left the writers with a sense of dread, such as the rain-drenched, inhospitable rocky shores of the northern coast of what became British Columbia, or what Parry (1828 Vol. 5, 34) described as the "motionless torpor" of the arctic winter: "an inanimate stillness … (for) … in the very silence is a deadness".

Unlike the native peoples, few European explorers knew how to survive in such environments. Not surprisingly this fear of such environments in Canada was linked to the survival theme found in many works of literature as Atwood (1972) has shown, yet is rarely acknowledged in scientific or ecological studies of environments (Draper 1998). However, the new literary approaches do more than show how aesthetics have influenced the choice of landscapes to describe and the feelings towards them. Other factors have also been shown to be influential, whilst the meanings behind particular portrayals must also be sought. Harley (1992) has shown these ideas in the context of map construction whilst the approach is well known in more visual art forms. For example, Feltes (1993) has shown that the canoe paintings of Frances Hopkins, long regarded as realistic paintings of canoes being used on the routes from which Canada was explored, encode a whole series of meanings. What Feltes has done is alert us to the meanings that lie behind the apparently simplistic realist portrayals. Certainly it can be argued that Hopkins's canoe painting can be 'read' on this literal plane. But the fact that the compositions seem to have been deliberately contrived to convey a series of 'other' meanings, linked to the power of the fur company officials being carried in boats manned by the métis, and even sexual tensions between the passengers, provides new understandings.

Towards the Construction and Understanding of Exploration Texts

The new interpretations by literary scholars have provided radical new insights into the way that the new environments were described by the early explorers. Rather than simply providing further examples of the deconstruction of exploration texts, this discussion summarises the variety of different influences that affect our understanding of exploration narratives (Figure 5.1). This is a shorter version of a more detailed provisional summary provided elsewhere (Davies 2003). Figure 5.1 shows the alternative ways in which our knowledge of the new environments can be derived from the writings of explorers. The simplistic approach (Route A in Figure 5.1) would be to assume that the public views, essentially our 'knowledge' of the new lands, are derived directly from some explorer's observations that are themselves a mimetic copy of what the new lands are like. A more detailed version of this approach would be to take the 1970s perceptual approach, which would assume that some images of the new lands are first derived or imposed and influence the way the text is written. This could be treated literally by the reading public, or privileged because they were the first descriptions (Route B). In practice, of course, the route by which we derive our knowledge of the new lands is far more complicated, for there are a myriad of factors involved in the construction of this knowledge, not all of which can be directly observed. In Figure 5.1, for analytical convenience, these are organised into four major categories of influence as: Context; Cognitive Construction; Writing Construction and Reader's Construction. However, there is little doubt that these factors overlap and interact with one another. In combination these various features show that it is impossible to accept that either our cognition, or our texts, or our understandings about the new environment, are produced through some mimetic reflection of the world. The rest of this section provides examples of these issues to illustrate the factors involved.

Context

All exploration narratives are situated within two background sets of influences. Figure 5.1 shows that the first is provided by the *stimulus of the external reality*, the other by the *context* of the exploration. In the former case the stimulus is never static, for the changing nature of this environment, especially the presence of storms or drought, can influence the views of the area and the results of exploration. Solomon

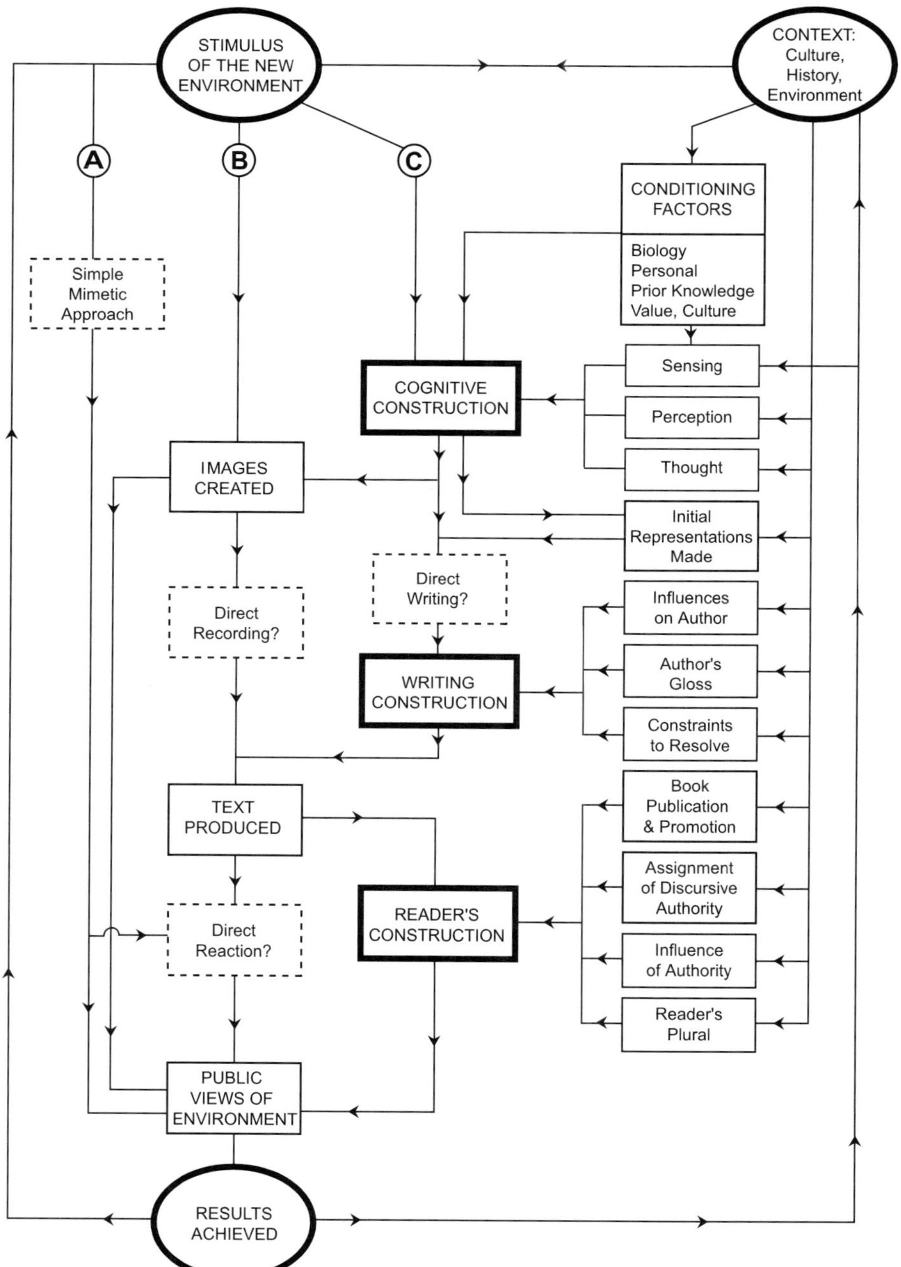

Figure 5.1 *The construction and interpretation of an exploration text.*

(2001), for example, argued that one of the undervalued reasons for the disastrous end of Scott's expedition to the Antarctic was that temperatures were colder than usual in the month before he died, so travel was more difficult. In Canada the expeditions of Palliser (1857–1860) and Macoun (1872) to the Prairies in the mid-nineteenth century, led to quite different opinions in relation to the suitability of the area for agriculture, the first negative the second positive, which may well be a product of different climate conditions. The second contextual influence is linked to the way that any explorer is a product of his or her own *culture, time and social environment,* all of which condition how the new lands are going to be seen and evaluated. Particularly important are the technological and political conditions of the day that influence how the environment can be used or traversed. For example, the Portuguese deliberately set out to collect maps of expeditions to help their exploration of the Atlantic (Russell 2000). But this was only a few generations after the Chinese deliberately withdrew from long distance expeditions, destroying most records of their voyages. Yet there is no doubt that they successfully reached the east coast of Africa in the early fifteenth century, and perhaps, according to Menzies's (2002) yet to be substantiated account, had also circumnavigated the world in 1421.

The emphasis upon conquest, wealth, or religious conversion by the sponsoring power, or the desire of the explorer to seek glory through his exploration, often obscures other influences on exploration. For example, Riffenburg (1993) has shown how many nineteenth century explorations were sponsored by rival newspapers that were 'creating', not simply 'reporting', news of the discoveries. It was linked to the hope that 'their explorers' would be the first to reach some destination, such as the North Pole or Central Africa, which would lead readers to buy their newspapers to find out the latest news and hence increase profits.

Cognitive Construction

The first of the major sets of influences that mediate between the stimulus of the new lands and the words of the explorer are essentially cognitive processes. Payne and Wegner (1998) have shown how experimental cognitive psychologists are expanding knowledge of how the mind works, which leads to influences far more complex than the basic ideas used in most perception geography. But this work is based on experiments in laboratories, and can give us only limited insights into how explorers from previous centuries derived their impressions of the environment. Nevertheless, the myriad of influences can be generalised into at least four major categories of factors that seem to account for how people build up their knowledge of the new environments they encounter.

One set of influences of the subsequent cognitive construction, summarized as *'conditioning factors'*, comprise those essentially personal reactions to stimuli. They include such features as biological endowment and personality issues, as well as factors such as prior knowledge and the values of individual explorers, although it is not always easy to tease out these last two categories of individual features from the societal background. In the former case there seems little doubt that many successful explorers showed a drive and determination that led them to succeed, whereas ordinary people would have given up. In addition, there now seems little doubt that gender does influence the way that the world is perceived, or at least how different features or perspectives are used by women explorers in their descriptions, as shown in studies by Blunt (1994), Blunt and Rose (1994) and Pratt (1992). The latter summarised the differences by suggesting that male narratives are usually dominated by "quests for achievement fuelled by fantasies of transformation and dominance ... (whereas) ... the exploratresses employ quests for self-realisation and fantasies of social harmony" (Pratt 1992, 168).

In the case of *prior knowledge,* the deficiencies in the historical record make it difficult to be conclusive about its influence. However there is strong circumstantial evidence that after failing to get the support he needed in the Spanish court to finance his own expedition across the Atlantic, John Cabot deliberately went to Bristol. Fishing expeditions from the port had probably found new fishing grounds across the Atlantic by the 1480s and had at least glimpsed a new land, although there were attempts to keep the former information secret (Wilson 1991). Moreover, Cabot found support from his successful 1497 expedition to the 'New Found Land' of Canada, not only from Henry VII, but also from prosperous local merchants such as Richard Ap Meryck, a major figure in Bristol who often signed his name as Americk (Davies 1986; 2003). His anglicised surname is so close to the name used to describe the new continent that it seems more than a coincidence and Bristol sources have long claimed that the continent was named after him (Hudd 1957), even though most authorities claim America was used to honour Amerigo Vespucci's expeditions. But perceived prior knowledge does not need to have a factual base. Most explorations in history took place in pre-literate society, with no records left for us to understand how new lands were discovered. Surviving information in folk stories that are orally transmitted through the generations is often heavily exaggerated and embedded in societal myth that may have foundation in fact.

Cognitive psychologists have identified a whole series of features that affect the way we perceive and think, many of which are undoubtedly linked together in complicated feed-back mechanisms. No one can experiment on long dead explorers, but the cognitive literature does provide concepts and clues to the way that different people recognise and select features on which to focus. The first of these relate to how we *sense* and experience our environment. In this context it is worth remembering Belyea's (1990, 13) observation that the overwhelming sensory experiences of explorers in the Canadian Arctic were of pain, cold, silence, fatigue and hunger. These sensations are often mentioned in exploration narratives, but the military tradition of many explorers meant that such deprivations were often downplayed in their books, since to complain was a sign of weakness. The result is that one can only glimpse at the effect of the journey upon the complete set of sensations, for the textual emphasis is usually on the visual. Another category of influences involves the conversion of the physical stimuli into internal representations in the brain, the production of what we often call images or *perception*. This involves a complicated series of processes ranging from the recognition, selection and memorising of information as well as its expression as language. But the way that the environment and the events in it are perceived and expressed varies by different cultures (Lowenthal 1961; 1967), which often produced misunderstandings between peoples. For example, Greenblatt (1991) has shown that on his third expedition Columbus was increasingly annoyed that the local inhabitants near Trinidad were reluctant to come close to his ships, for he wanted to trade with them. He decided to show his peaceful intentions by getting his seaman to sing and dance, as if in a party. However, such behaviour was interpreted as hostile acts by the natives, who soon unleashed a series of arrows on the unsuspecting Spaniards. In native cultures the type of posturing and noises they 'saw and heard' were interpreted as the type of the provocative show displays used by many indigenous groups to show their bravery before a battle is joined.

The next category of influences are the processes that psychologists summarise as *thought,* which include such features as decision making, conceptualising, and cognitive skills. Among the major factors that make up the process we describe as thought, there seems little doubt that the way that decisions are arrived at may have a huge impact on the success of any venture. Allen (1975b) has suggested that the best explorers were those who quickly modified their routes when new information was obtained, instead of sticking to pre-conceived plans that were usually based on inaccurate speculation. In terms of cognitive skills we also know that many other species have far more developed way-finding ability than our own, although much of this may be innate rather than conscious. However, many Aboriginal Peoples, living closer to the environment, do have senses more finely tuned, or trained, to recognise environmental clues that escape Western man, enabling them to find a way across apparently unknown or trackless areas. Turnbull (2000) argued that Polynesian navigators had intimate knowledge of star systems and of how far they had travelled each day, but they were also able to sense the presence of islands from wave patterns, long before they could be seen. In addition, he observed that they constantly modified their voyage almost by instinct so that their voyages were based on what he called a 'performance' rather than a 'representational' navigation linked to some latitude–longitude chart. There are few similar authoritative studies in the literature about Canadian Aboriginal Peoples but there seems little doubt that many had similar way-finding ability; the Inuit in particular had great skills that enabled them to live in and traverse one of the most difficult environments in the world.

Writing Construction

Once explorers began to write down their accounts of their expeditions another set of influences must be considered as affecting the way that the narratives are written, for again it is clear that there these cannot be mimetic records of what was seen. The first stage of the writing process consists of the *initial representations* that were created, often in the form of notes or a daily log. Often written under conditions of environmental discomfort or hazard, they are rarely examples of enduring prose. But many scholars have shown that the daily log often differed significantly from the final journal. Some of the changes were undoubtedly the result of new information, or discoveries by the explorers as they progressed through the unknown land. But it can also be shown that there were many cases of re-writings at a later date when the final narrative was compiled. Some were a product of second thoughts, or by subsequent reading. But it is also clear that explorers employed editors or ghost writers who saw the need to provide more vivid descriptions. These revisions were often written with a deliberate concern for the market, for the people who would buy and read the books. Davis's (1991; 1995; 1998) careful comparison of Franklin's initial notes, his daily logs of his two expeditions to the Arctic shores between 1819 and 1827 that were lodged with the Admiralty, and his final narrative illustrates the major differences between these texts. He showed that Franklin added pejorative adjectives, such as the word 'grotesque' in his final narrative, to describe native dances, presumably to emphasize the 'savage' nature of native peoples that Dickason (1997) has shown to be such an unjustified characteristic of European descriptions of native peoples in Canada. Such examples show that many of these early exploration narratives were not the dispassionate, rational descriptions of the new lands and their peoples that many have assumed.

The second category of features that affect the writing of the exploration narrative consists of a series of *influences on the writer*, some of which relate to the cultural and historical context and others to the individual. All writers of exploration texts are affected by previous works, opinions and the culture in which they live as well as their own motivations and ability to raise the money to finance the expedition, or to gain the prize of being the leader chosen by the sustaining power. Many exploration accounts were written for the sponsor, such as the monarch, not for the general public and often reflected the desire to please this particular reader and his court. By contrast many other explorers seemed, from the first, to aim their accounts deliberately at the general reader. Although the sensational style became most obvious in the nineteenth century, an early example can be seen in James's narrative of his abortive voyage to find the elusive North West Passage (James 1633). The very title – *The Strange and Dangerous Voyage of Captaine Thomas James in his intended Discovery of the Northwest Passage into the South Sea* – gives his intentions away, that of concentrating upon the difficulties of the quest, upon his journey, not on what was found.

Once the explorer begins to actually write his text another series of writing strategies can be identified which again cast doubt about the presence of any simple one-to-one relationships between the physical stimuli and the final text, influences which can be summarised as the *author's gloss*. These include features such as: the general organisation of the narrative; the insights and expressions used; the writing strategies – such as style, language choice, the rhetorical devices adopted to make the text interesting to the reader; and the use of other forms of representations, such as maps or sketches; as well as the meanings encoded. However, the elaborate use of metaphors based on classical or biblical themes fell into disfavour from the seventeenth century onwards. Some of these issues relating to the way that many explorers were significantly influenced by the aesthetic ideas of the day have already been described above. But Pratt (1992) has shown that two other rhetorical devices were also used to create what she called additional value for the discoveries in the eyes of the readers. One was described as "density of meaning" in which a variety of modifiers are used to give enhanced meaning to any description. James's textual descriptions of ice conditions can be used to illustrate the point. Unlike his rival Foxe (1635), who was also in Hudson Bay on the same quest in the summer of 1632, James did not simply use the word 'ice' by itself. It was always "unusually large", "as high as our mast head", "towering", "twice as high as our mast head", whilst later on he makes an appeal to other senses by describing the ice floes as creating a "hideous noise" (Davies 2003, 229). One need not doubt that the ice frequently threatened to overwhelm James and his crew; but the use of such descriptors certainly added to the element of fear that he wanted to create, and improved the readability of the narrative. Pratt (1992) also identified the category of "mastery" in many explorer's descriptions, namely the way they frequently viewed the new lands from a position of superiority and which often led to descriptions about how the area could be improved by European progress. But not all explorers were in a position of superiority. It is worth remembering that many explorers depended on native people to take them to the places that they were then supposed to have 'discovered'. For example, Venema (2000) has described how Samuel Hearne was really a subordinate to Chief Matonabbee who was the real leader of the journey to the Coppermine river in 1771, whilst his descriptions acknowledged the vital role played by women in finding edible plants and acting as pack carriers in the annual migrations. Finally, there is little doubt that the texts of the most successful writers were improved by the use of insightful phrases that provide succinct and lasting impressions of the land being traversed. Cartier's description of the forbidding Quebec north shore as "the land God gave to Cain" has resonated through the ages, portraying the Canadian Shield, or at least its agricultural capacity, as a land only fit for those that God had cast out of his sight. These issues show that although Atwood's emphasis on survival in a difficult environment is a basic and characteristic theme in early Canadian literature, it must be seen as only *one* interpretation of the environment.

The fourth category of influences on the writer consists of the *constraints* that are faced. Many explorers had sponsors they could not afford to alienate, so their texts often reflect this fact. In addition, the relative ease of publication today makes it difficult to appreciate the difficulty that many explorers had in getting their work published, especially those who remained in the distant lands they explored. This had a huge effect upon the dissemination of knowledge of the results of their explorations. One of the best examples may lie in the case of David Thompson, the master mapmaker of the Canadian continental interior and the northern fringes of what became the U.S.A. He was the son of Welsh-speaking immigrants to London, who went to the Hudson Bay Company in western Canada as a youth in 1784 and never returned to Britain. His work is far less known than his contemporaries, Lewis and Clark, leaders of a major expedition through the northern Great Plains and across the mountain barriers to the Pacific in the first decade of the nineteenth century (Allen 1975a). Their journey was sponsored by an American government with its own dream of western expansion; moreover, the expedition produced narratives that were extensively edited to improve their readability. By comparison, Thompson worked for fur companies interested in their own commercial agendas. His early nineteenth century map of the continental interior of Canada and northern U.S.A. was one of the most accurate of its day. But it was company property and was never published, languishing, but still used, in one of the major depots of his company. Fortunately it was rescued decades later and now hangs in the entrance hall of the Ontario Archives.

Readers' Construction

The last category of influences relates to the derivation of meaning by readers from the various exploration texts. The realist approach to landscape description by geographers and environmentalists has led to the assumption that there is one correct or perhaps best interpretation, that may be modified as new information becomes available or the situation changes. However, students of prose works in particular have always argued that there can be many interpretations of any written text, depending upon the perspective taken, or the issues stressed, a relativist approach that has been given new impetus by the deconstruction ideas of French critics. These ideas mean that exploration narratives also need to be viewed through the prism of a *readership filter*. We take or accept different pieces of information from books, and weigh them differentially, often under the influence of major critics or scholars in the field, which obviously affects the knowledge that we derive.

The most obvious influence upon 'reader knowledge' comes from the fact that not all exploration narratives were published. Many were withheld as commercial secrets by their sponsors or deliberately repressed as state policy. Again the case of David Thompson can be used to illustrate the issue. It can be argued that he was far less known than his contemporaries simply because he never had the contacts to get his fascinating journal published, for he died in poverty in 1857. Thompson's achievements only became part of the public domain of information a half century later through edited versions of his narrative (Tyrrell 1916; Glover 1962).

Even if books are published their influence is often dependent upon the way that publishers print and promote the work. However, as important is the way that the opinions about the books and the utility of the descriptions are influenced by what can be called the voices of authority, such as previous authors in the field, or critics. Their reviews provide an authority structure that evaluates the 'knowledge' contained in a book, but may well distort the intent of the book for their own purposes, with negative effects on subsequent readers. A good example of this problem may be seen in the search for the North West Passage (Savours 1999), especially James's voyage and wintering in Hudson Bay between May 1631 and October 1632 that was described in his book (James 1633). Most of the references to James's work in the last two centuries are dismissive, seemingly taking their lead from the negative opinions of Barrow, the influential second secretary of the British Admiralty. In his book on Arctic expeditions Barrow (1818) did praise the quality of James's writing, but argued that James was a poor seaman because he was always on the brink of disaster. James did not discover a great deal, other than the southern shore of Hudson Bay and its southern extension in the bay that he named after himself. But we now know that his task was impossible, given that his instructions were to seek the passage *through* Hudson Bay. Davies (2003) has recently provided a re-assessment of James's book noting that it was initially quoted with approval by the major compilers of voyages of discovery.

Also, James's lengthy comments about the effects of cold were extensively quoted in the works of Robert Boyle, Britain's premier scientist in the late seventeenth century (1774). Moreover, James seems to have been given little credit for the remarkably accurate descriptions of his longitude and his environmental insights. In addition, James's work crossed the scientific-artistic divide, for the poet Samuel Coleridge seems to have adopted many of James's images into his famous poem *The Ancient Mariner* (Davies 2003; Rose 1970). So there seems little doubt that James's work had what can be called 'instrumental value' beyond itself, since it inspired subsequent work. So why was Barrow so negative? The reason seems to lie in the fact that Barrow was busy planning new expeditions of discovery to the Arctic and other areas after the Napoleonic wars, using surplus British naval vessels. So James's conclusion that there was no North West Passage, or rather not one that was commercially feasible, given ice conditions, was not a welcome conclusion. In addition, the new laconic style of the new breed of explorers and British naval officers, whose own prose ignored or played down danger, probably found James's constant repetition of his constant brushes with potential disasters unpalatable. It is from this time James's work was increasingly dismissed; he is given a poor 'press' in most standard sources today, for critics focus upon the 'product' of the exploration, the abortive quest, not the ordeal that the voyage turned into. Yet it was this fast-paced tale of heroism, tribulation and adventure in what was an alien environment to a British readership, that enthralled James's contemporaries and readers in subsequent centuries, making it such a best seller in its day. Probably James deliberately wrote his narrative in this way, simply because he did not find the much sought after passage. So to read the narrative *only* for its value as contributing to the some 'exploration result' seriously misrepresents James's intent. James may not have been an explorer of the first rank but there seems little doubt that he was a master seaman and leader, an early environmentalist because of his observations, as well as a master author who could enthral his readers.

Finally, the derivation of what might be called 'public knowledge' from any exploration text is also dependent upon the previous information base, values and character of readers. Moreover, readers may gain opinions or attitudes from what they get out of a book, such as the derivation of new information, pleasure, fear or other emotions, or from the way an exploration narrative may consciously or unconsciously represent a particular ideology, especially that of the right of conquest. Whether these attributes are shared by others or not is always a moot point. What seems undeniable is the presence of what is usually described as the 'reader's plural' – the possibility of many different views about the value of a book, and what it represents. Certainly this may lead to the chaos of alternative opinions. But it has already been argued that these plural views are not necessarily a set of unique opinions; within certain groups there is often a more or less shared consensus, at least in one time and place. But opinions change and we must always be careful of assuming that a narrative of exploration has only one interpretation, for personal judgment is involved. For example, Belyea (1990) criticized MacLaren's (1984b; 1985b) discussion of the way that Franklin's narratives of his Canadian explorations adopted the aesthetics of the picturesque or the sublime in their descriptions. She noted that Franklin *also* selected and discussed other features that could not be attributed to these traditions, producing a double discourse. Belyea is emphatic that her 'reading' is different from that of MacLaren:

> The illustrations and text of Franklin's published Narrative record the expedition's achievements despite all odds, even aesthetic ones: this is MacLaren's reading. But I prefer to think of the Narrative as patching together several discourses and exploring European conventions as well as the deserts of America. (Belyea 1990, 24)

The result of all these influences means that exploration narratives, or even texts in general, can never be seen as static representations. Just as the words on the page are never a simple mimetic representation of what was seen by the author, so opinions about the value or utility of a book will change through time. It is the presence of different attitudes, beliefs and ideologies that provide a reader's construction of meaning about any text – a set of influences to set alongside the constructions linked to the initial cognition and subsequent writing. James's narrative of his exploration and wintering provides a particularly good case study of the changes in the way that his work was received, initially with enthusiasm, but with scepticism from the early nineteenth century. Opinions of James's significance in exploration or in exploration writing – not necessarily the same thing – are themselves 'constructed' by his critics. They may be reading the text from a particular point of view, or from opinions expressed by others. But few exploration texts were 'read' in such ways until two decades ago. The texts were regarded as realist observations, useful for what they told about the discoveries, not the underlying values and meanings that infused the words and structures used by the authors.

Conclusions

This chapter began with a quotation from Wreford Watson that emphasised that our knowledge of places depends on 'how' we see the world, not on the reality of 'what' may be there. But the new perspectives from literary criticism that have been applied to exploration narratives have shown that this idea needs to be extended. *Exploration narratives can no longer be seen as 'objective' descriptions of sights* that were written down. The realist assumption has been destroyed and the books written by explorers must be appreciated as *constructed narratives influenced by many factors.* Not only do these tell us as much about the aesthetic preferences and intentions of the explorer and his culture and context, as what was in the environment, but they also illustrate that there may be very *'different readings'* of any text which produce very different contributions to what we accept as knowledge. So the writings of explorers in Canada and other areas are now being treated as examples of *alternative ways of representation and reading*, rather than being primarily seen as *way-finding* through the discovery and description of new lands. In addition, although this essay has focused upon the first descriptions of the new land by explorers, the ideas described here apply to all sorts of texts that are written about the environment, whether scientific, ecological, literary or others, as well as to the opinions that readers derive from the works – part of the search for meanings that lie behind the written words and their construction. In our study of Canadian environments we need to interrelate these domains to obtain a more comprehensive approach. Understanding the way we construct our knowledge, together with a clearer appreciation of the various influences that affect it, may provide a way of achieving this end.

References

Allen, J. L. (1975a). *Passage Through the Garden: Lewis and Clark and images of the American West.* Urbana: University of Illinois.

Allen, J. L. (1975b). Lands of myth, waters of wonder: the place of the imagination in the history of geographical imagination. In D. Lowenthal and M. J. Bowden (Eds.) *Geographies of the Mind: Essays in historical geography.* Oxford: Oxford University Press.

Atwood, M. (1972). *Survival: A thematic guide to Canadian literature.* Toronto: Anansi Press.

Atwood, M. (1995). *Strange Things: The malevolent North in Canadian Literature.* Oxford: Clarendon Press.

Barnes, T. J., and Duncan, J. S. (1992). Introduction: writing worlds. In T. J. Barnes and J. S. Duncan (Eds.) *Writing Worlds: Discourse, text, and metaphors in the representation of landscape.* London: Routledge.

Barrow, J. (1971 [1818]). *A Chronological History of Voyages into the Arctic Regions.* London: John Murray, 1818. Reprinted C. Lloyd (Ed.) New York: Barnes and Noble, 1971.

Belsey, C. (2002). *Critical Practice.* London: Methuen.

Belyea, B. (1990). Captain Franklin in search of the picturesque, *Essays in Canadian Writing,* 40, 1–24.

Blunt, A. and Rose, G. (Eds.) (1994). *Writing Women and Space.* New York: Guildford Press.

Blunt, A. (1994). *Travel, Gender and Imperialism.* New York: Guildford.

Bonnycastle, S. (1991). *In Search of Authority: An introductory guide to literary theory.* Peterborough, Ontario: Broadview Press.

Boyle, R. (1774). New experiments and observations touching cold. In D. McKie (Ed.) *The Works of the Honorable Robert Boyle* edited by W. Birch. London: 1774. (New edition 1965), Vol. 2, Hildesheim: G. Olms.

Campbell, M. B. (1988). *The Witness and the Other Word: Exotic European travel writing: 400–1600.* Ithaca, New York: Cornell University Press.

Clifford, J. (1986). Introduction. In J. Clifford and G. Marcus (Eds.) *Writing Culture: The poetics and politics of ethnography.* Berkeley: University of California.

Davies, W. K. D. (1986). The Welsh in Canada: A geographical overview. In M. Chamberlain (Ed.) *The Welsh in Canada.* (Reprinted 2001). Swansea: Canadian Studies In Wales Group, University of Wales Swansea.

Davies, W. K. D. (2003). *Writing Geographical Exploration.* Calgary: University of Calgary Press.

Davies, W. K. D. and Gilmartin, M. (2002). Geography as a cultural field. In R. Gerber and M. Williams (Eds.) *Geography, Culture and Education.* Dordrecht: Kluwer.

Davis, R. C. (1991). History or His/Story?: The explorer cum author, *Studies in Canadian Literature,* 16, 93–111.

Davis, R. C. (1995). *Sir John Franklin: The first Arctic expedition.* Toronto: Champlain Society.

Davis, R. C. (1998). *Sir John Franklin's Journals and Correspondence: The second Arctic land expedition, 1825–27.* Toronto: The Champlain Society.

Dickason, O. P. (1997). *The Myth of the Savage*. Edmonton: University of Alberta Press.
Draper, D. (1998). *Our Environment: A Canadian perspective*. Toronto: Nelson.
Dubinsky, K. (1998). The pleasure is exquisite but violent: The imaginary geography of Niagara Falls in the nineteenth century, *Journal of Canadian Studies,* 33, 64–65.
Eagleton, T. (1983). *Literary Theory: An introduction*. Minneapolis: University of Minnesota Press.
Feltes, N. N. (1993). Yoy(ag)euse: Gender and gaze in Frances Anne Hopkins's canoe paintings, *Ariel,* 24, 7–20.
Foxe, L. (1635). *Northwest Foxe*. London: Alsop and Fawcet. Reprinted (1965) New York: S. R. Publishers: Johnson Reprint Corp.
Glickman, S. (1998). *The Picturesque and the Sublime: A poetics of the Canadian landscape*. Montreal and Kingston: McGill-Queen's Press.
Glover, T. (1962). *David Thompson's Narrative: 1784–1812*. Toronto: The Champlain Society.
Greenblatt, S. J. (1991). *Marvellous Possessions*. Chicago: University of Chicago Press.
Gregory, D. and Duncan, J. S. (1999). *Writes of Passage: Reading travel writing*. London: Routledge.
Harley, J. B. (1992). Deconstructing the map. In T. J. Barnes and J. S. Duncan (Eds.) *Writing Worlds: Discourse, text, and metaphors in the representation of landscape*. London: Routledge.
Harris, J. (1705). *Navigantium atque itinerantium bibliotecha: A Compleat collection of voyages and travels*. Two Volumes. London: T. Bennett. Revised and enlarged by J. Campbell, (Ed.). London: Reprint. London: T. Woodward *et al.* 1744–48. New edition, 1764.
Hudd, A. E. (1957). Richard Ameryk and the name America. In H. P. R. Finberg (Ed.) *Gloucestershire Studies*. Leicester: University of Leicester Press,
James, T. (1633). *The Strange and Dangerous Voyage of Captain Thomas James*. London: John Legatt for John Partridge. Reprinted (1973). Toronto: Coles.
Kreisel, H. (1968). The prairie: A state of mind. *Transactions of The Royal Society of Canada*, 6, 169–173.
Lowenthal, D. (1961). Geography, experience and imagination, *Annals of Association of American Geographers*, 51, 241–260.
Lowenthal, D. (1967). *Environmental Perception and Behaviour*. Chicago: University of Chicago Press.
MacLaren. I. S. (1984a). Samuel Hearne and the Landscapes of Discovery, *Canadian Literature,* 103, 27–40.
MacLaren. I. S. (1984b). Arctic exploration and Milton's 'Frozen Continent', *Notes and Que*ries, New Series, 31, 325–326.
MacLaren. I. S. (1985a). Aesthetic mapping of nature of the west by the Palliser and Hind survey expedition, *Studies in Canadian Literature*, 18, 24-52.
MacLaren. I. S. (1985b). The aesthetic mapping of nature in the second Franklin expedition, *Journal of Canadian Studies,* 20, 39–57.
MacLulich, T. D. (1979). Canadian exploration as literature, *Canadian Literature,* 81, 72–85.
Menzies, G. (2002). *1421: The year China discovered the world*. London: Bantam.
New, W. H. (1997). *Borderlands: How we talk about Canada*. Vancouver: UBC Press.
Parry, W . E. (1828) *Journals of the First, Second and Third Voyages for the Discovery of a North West Passage from the Atlantic to the Pacific in 1819–25. Vols 1–5*. London: John Murray.
Payne, D. G. and Wenger, M. J. (1998). *Cognitive Psychology*. Boston: Houghton Mifflin Company.
Pinker, S. (1997). *How the Mind Works*. London: Penguin.
Pratt, M. L. (1992). *Imperial Eyes: Travel Writing and Transculturation*. London: Routledge.
Ricou. L. (1973). *Vertical Man in a Historical World*. Vancouver: University of British Columbia Press.
Riffenburgh, B. (1993). *The Myth of the Explorer.* Oxford: Oxford University Press.
Rose, M. (Ed.) (1878). *The Rime of the Ancient Mariner: Illustrations by G.Doré*. Reprint (1970). New York: Dover Press.
Russell, P. (2000). *Prince Henry The Navigator.* New Haven: Yale University Press.
Savours, A. (1999). *The Search for a North West Passage*. London: Chatham.
Shields, R. (1991). *Places on the Margin*. London: Routledge.
Simpson-Housley, P. and Mallory, W. (1986). *Geography and Literature: A meeting of disciplines.* Syracuse, New York: Syracuse University Press.

Solomon, S. (2001). *The Coldest March: Scott's Antarctic expedition*. New Haven: Yale University Press.

Turnbull, D. (2000). *Masons, Tricksters and Cartographers.* Amsterdam: Harwood.

Tyrrell, J. B. (1916). *David Thompson's Narrative of his Explorations*. Toronto: The Champlain Society. Republished (1960). New York: Greenwood Press.

Venema, K. (2000). Under the protection of a principal man, *Canadian Writing*, 70, 162–190.

Warkentin, J. (Ed.) (1964). *The Western Interior of Canada*. Toronto: McClelland and Stewart.

Warkentin, G. (Ed.) (1993). *Canadian Exploration Literature: An anthology*. Toronto and Oxford: Oxford University Press.

Watson, J. W. (1969). The role of illusion in North American geography, *Canadian Geographer,* 13, 10–27.

Watson, J. W. (1983). The soul of geography. *Transactions of the Institute of British Geographers, New Series*, 8, 385–399.

Whorf, B. (1956). *Language, Thought and Reality.* Cambridge, Mass: MIT Press.

Wilson, I. (1991). *The Columbus Myth: Did men of Bristol reach America before Columbus?* London: Simon and Schuster.

6. CHANGING PUBLIC PARTICIPATION AND THE ENVIRONMENT OF SWANSEA EAST

GRAHAM HUMPHRYS

Introduction

The last third of the twentieth century saw the growth of a continuing debate on the evolution and governance of liberal democracy (Paehlke and Torgeson 1990; Held 1996; Dryzek and Schlosberg 1998). The debate is closely linked with the growing public interest in environmental change and environmental management and especially with environmental sustainability. This is because much of the interest is related to how democratic government can properly accommodate the disparate interests involved in changing the environment of individuals as well as the environment at national and global levels. In the United Kingdom especially there has been a long-standing public attitude of deference to the presentation and representation of their views and interests by elected politicians, and an acceptance of the political decisions made on the advice of technical professionals within the public administration system. Before 1960 this acceptance was largely justified by the perceived competence and integrity of those involved. What subsequently changed was an increasing complexity of environmental issues, accompanied by an increase in environmental regulation, reduced deference by a more educated public and a growing awareness of the significance of the environment in the lives of individual electors. There was also growing recognition that different groups and individuals experience, observe and interpret environments in different ways depending upon their own interests, values and backgrounds. Each will see the environment differently and their perspectives will determine the ways they choose to present and represent the environment. It will also determine their choice of what they perceive as the most important challenges within the environment, and how they prefer them to be dealt with. Any 'truth' about the environment derived from objective analysis and understanding of the processes at work will thus be selectively used and presented, though it is not always in the interest of individual presenters to acknowledge their selectivity explicitly even if they are aware of it.

Acknowledgement of the subjectivity of much of the environmental debate and of the role of selective interpretation in the presentation of environmental knowledge

G. Humphrys and M. Williams (eds.), Presenting and Representing Environments, 89-104.

has resulted in a more frequent rejection of actions affecting the environment which are proposed by public bodies and by private companies. The effect of uncritical attitudes in the past was to advantage those who wielded power and had access to the necessary knowledge. This allowed them to use and sometimes even abuse the environment for their advantage and to the disadvantage of others and at the same time claim that they were doing right or had no alternative. From a contemporary perspective, it is possible to look back and see how important were the presentations and representations of the environment by those with power in determining the shaping of any one environment over time. All this has given rise to a recognition that the validity of the inherited model of presentation and representation of the environment in liberal democracy, and the place of public participation within it, at the very least needs to be reviewed. In summary, the environment has raised challenges for government and governance that are still being worked out in the political arena.

The theme of the role of direct public participation as an element of true liberal democracy is raised initially in this chapter and then illustrated in an examination of the evolution of environmental change in Swansea East as defined below. This inner urban area has been chosen because it experienced major environmental change after 1960 which still continues, and where new environmental issues have been raised which are yet to be resolved. It is also useful because it is an example of urban rather than rural environmental change and can thus be more readily related to by urban dwellers elsewhere who now form the majority of the world population. Indeed, much of the debate on environmental issues has been concerned with the abuse of the rural environment. Less attention has been paid to dealing with the environmental challenges of older industrial and urban areas. In these, the restoration of land devastated by past industrial and urban activities is more important and different pressures arise from the use of the environment within the urban fabric. The inherited landscape of urban areas is an important part of the local social capital and an integral part of the quality of urban life. It is now seen as belonging to, and for the use and benefit of, the community and others who refuse to continue the resigned acquiescence of their forefathers.

Liberal Democracy and Public Participation

Public participation is generally considered an essential ingredient of good liberal democratic government. Indeed, without it liberal democratic government could be said not to exist. "The extent to which there should be public participation, and how it should occur, goes to the heart of a nation's political values, its concept of the state and the state's relationship with its citizens, and its concept of how public business is properly carried out" (Barton 2002, 77). However, it is quite easy to see how some political theorists have argued for democratic elitism as a suitable form of modern government (Held 1996). The complexity of the modern world is such that well qualified officers are needed in government who have the skills necessary to advise elected policy makers when they are dealing with the problems of society. By definition, the level of skill needed means that these advisors form a meritocracy. This is particularly true in the environment field where high-level scientific and technical understanding is needed. However, this does tend to result in a dominance of scientific rationalism when the environment is dealt with, which, as has been argued in the introduction to this volume, is usually accompanied by and privileges some value judgements often to the exclusion of others. Experts have their prejudices, too, even if they remain unacknowledged.

It follows that in such a system the views of this meritocratic elite usually prevail and that these do not necessarily achieve a satisfactory presentation and representation of the views of the majority of the electorate. A bureaucratic elite of this sort is likely to be efficient, though whether it will also be effective will depend upon the standpoint of their assessor. Barton suggests that, "An elite has little enthusiasm for public participation. Even a democratic form of elitism based on rationality and official expertise, sees no need for public participation in solving problems, it knows the answers itself" (2002, 85). Other problems with having a dominant meritocratic elite decide matters affecting the environment came to the fore in the last part of the twentieth century. They emerged from a growing recognition of the limitations of the scientific knowledge and understanding of the environment. This led to the suggestion that a post-normal science approach was more sensible (Funtowicz and Ravetz 1993; Ravetz 1999; Muller 2003). Essentially, this provides a way of deciding what is the best action to take when that action has environmental implications, when there is uncertainty in the knowledge available and when there are various differing interests involved in the decision making process (Hulme and Turnpenny 2004). Input of knowledge and process understanding is normally provided by objective peer-reviewed environmental science and by consultancies. The former is only partial because of the complexity and constant change of the environmental systems researched, while the reports of the latter are influenced by the remit of the clients who pay for them. The elected representatives, who are the users of the information to take the decisions, are in turn limited by the extent of their understanding and influenced by the representations of interested parties who often have opposing views. All of this is affected by the limitations on time and funds available. The post-normal science approach accepts all of this, but recognises that decisions on action must be taken to deal with particular environment questions (Ravetz 1999). This emergent realism is growing in acceptance in the early twenty-first century.

Where the officers of a democratically elected government exercise undue influence on the decisions made by the elected representatives it can be argued that there is evidence of democratic elitism. In a mature liberal democracy those who take the determining decisions are the elected representatives. Elections provide the safeguard of their accountability to the electorate, with the bureaucracy held accountable through the decisions of the elected or the person, often a Minister, who is responsible for it. Even in this there is also the danger of democratic tyranny where the majority consistently takes decisions for its benefit to the detriment of the minority of the electorate who did not vote for them. There is also the temptation among the elected to 'know the interests of others' who through blindness or ignorance cannot perceive their interests for themselves. For the system to work at its optimum, the presentation and representation of the public interest should be supplemented by public participation through the direct involvement of people or groups. It is this that has been receiving far greater attention since 1960, partly stimulated by environmental activists. What such participation does is allow for other values and popular opinion to be presented and represented when decisions are being taken, and non-quantifiable preferences derived from outside the rigid constraints of scientific rationalism to at least be considered and evaluated. It is, of course, easier to propose this as the right way forward for liberal democracy than to devise a practical system for it to be successful. Greater participation by the public in political matters, even in elections where it is voluntary, is proving hard to achieve in most liberal democracies in the twenty-first century. There are good reasons for this. Participation makes a demand on people's time and often on other resources, and many lack the experience or the expertise to develop the confidence to participate. Others refer to a 'culture of contentment' where the majority are broadly satisfied with the *status quo* as if to justify apathy. In such a situation participation is often only by those who have a particular personal interest, which may not be that of the majority or even a sizeable minority, or by special or single interest groups seeking to promote their cause. For good reason the way forward for good presentation and representation is still being explored.

The Swansea East Environment Inherited in 1960

Most of Swansea East, as dealt with here, is the south and west section of the modern parliamentary constituency of the same name. It lies at the heart of Swansea Bay City, a conurbation of some 400,000 people located on the coast at the western end of industrial South Wales (Humphrys 1990). Within it, the urban areas form an L shape (Figure 6.1). The vertical element of the L, some six kilometres from north to south, is known as the Lower Swansea Valley. It comprises the valley floor of the lower reaches of the River Tawe as far as the sea, together with the adjacent eastern valley sides. The horizontal element of the L, extending about five kilometres east-west, is the urbanised area bordering the sea to the east of the river mouth. Here it extends about a kilometre beyond the Swansea City and Swansea East Constituency boundary into the neighbouring County of Neath-Port Talbot. Between the two elements is Kilvey Hill rising to nearly two hundred metres with a covering of heath on its steeper slopes and mixed woodland extending over the summit. Immediately east of Kilvey Hill is Crumlyn Bog, a large area of marshland extending three kilometres inland from the coastal strip and including a designated Site of Special Scientific Interest.

Coal mining on a small scale was already well established on the east side of the Lower Swansea Valley by 1700, with the coal being exported from the small port of Swansea at the river mouth. Soon after this the first copper smelter was established on the floor of the valley at Landore and by 1823 there were eight other copper smelters operating in the Lower Swansea Valley (Williams 1940). By then the

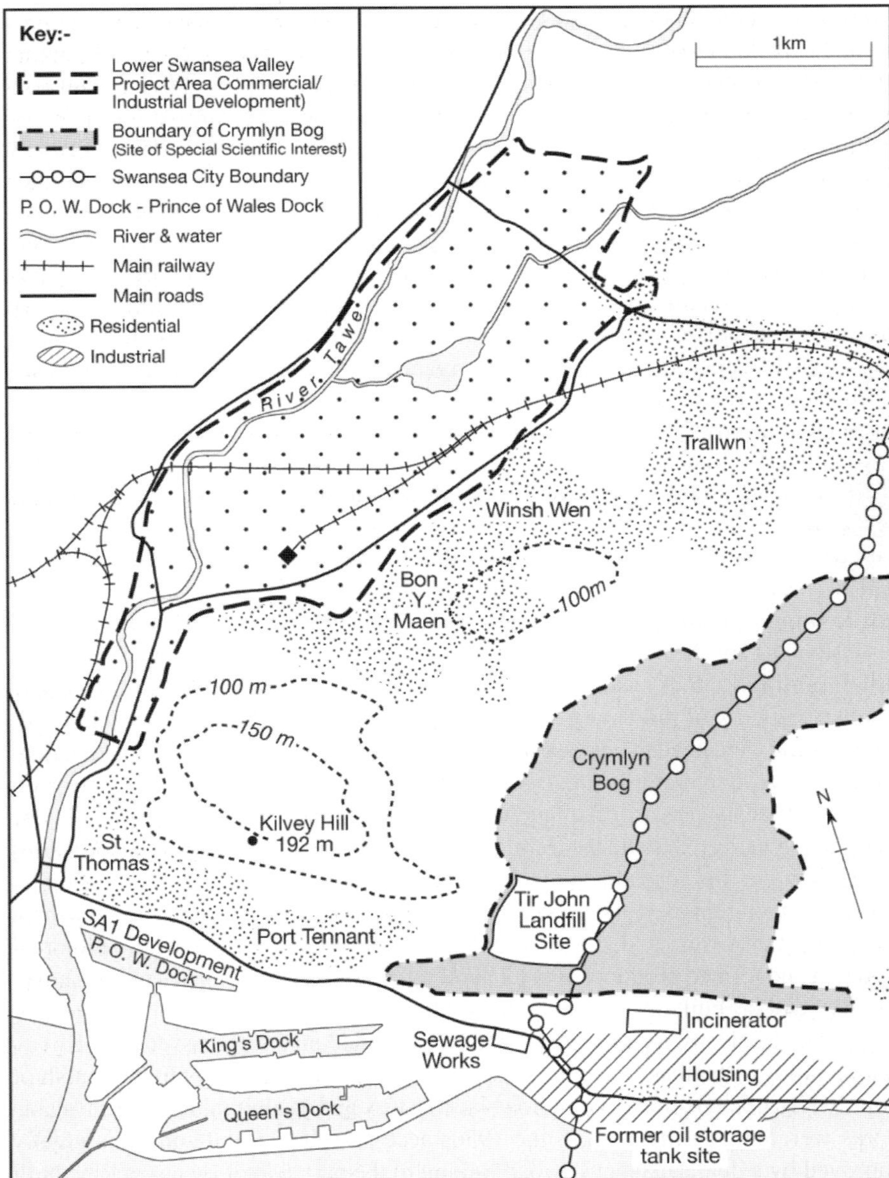

Figure 6.1 *A map of Swansea East: principal features.*

Swansea region was smelting over half of the world output of copper (Hughes 2000). Around this early industry a further complex of non-ferrous smelting works developed which produced copper, lead, zinc, gold, silver, nickel and cobalt. Industrial expansion continued throughout the nineteenth century, with the addition of a range of other heavy industries of which iron, steel and tinplate making and the manufacture of coal briquettes were particularly important. For the most part, the industrial works occupied the flat valley floor near the River Tawe, with housing for the workers built nearby on the lower valley sides. Waste from the industries was dumped on sites adjacent to the works to cover almost all the rest of the valley floor and it spread up the valley sides in places. The environmental devastation was not confined to the land. The smelting process generated large quantities of smoke and fumes consisting of sulphur dioxide, sulphurous and particulate sulphuric acid. With the prevailing wind from the southwest, these fumes blew over the eastern slopes of the valley resulting in almost complete destruction of the vegetation in places and affecting crops and livestock. Local rivers and streams were poisoned as toxic chemicals leached out of the waste tips by the rainfall drained into them. As early as the 1920s the industrial activity in the valley had begun to decline. Works were abandoned and their derelict buildings added to what was now a complex industrial wasteland. Much of the environment had been used and abused and then abandoned. By 1960 the devastated area covered 300 hectares, mostly waste tips of various toxicity together with derelict buildings and a few surviving working factories (Hilton 1967).

The coastal zone developed differently. As early as 1824 a canal was built from the east to bring coal and iron from the Neath valley for export to the natural harbour here. Housing for the port workers was built on the adjacent lower slopes of Kilvey Hill. From 1854 a dock system developed around the mouth of the Tawe, linked by a complex network of railways to the industrial hinterland across the coalfield to the north. These were needed because the tidal range in Swansea Bay is among the highest in the world, averaging in excess of eleven metres. As further docks were built seaward, extensive areas of flat land were created around them and used for railway marshalling yards to handle the trainloads of coal arriving for export. The industries that grew up in this coastal zone included tinplate making, coal-briquette manufacture and wagon repairs, none of which generated large waste tips. In the 1920s large oil storage tanks were added to the landscape east of the docks to handle imported crude oil and refined products for a local oil refinery. Local Authority (public) housing extended the built up area along the south side of Kilvey Hill after 1930 (Herbert 1971). A large electricity generating station was added on the eastern edge of Kilvey Hill in 1940, with an associated fly ash waste tip extending east onto the marsh of Crumlyn Bog. Soon after this a carbon black factory and an aluminium wire and cable works were built in the area. In the 1950s access to the coastal zone was greatly improved by a dual carriageway road linking to the east side of Swansea Bay, built along a line separating the docks and the housing.

Changing Environments after 1960

The Lower Swansea Valley

In 1960 the environmental challenge presented by the Lower Swansea Valley was how to transform the 300 hectares of devastated and mostly derelict industrial land. With a lead from the local university, the Lower Swansea Valley Project was established in 1962 as a joint venture between the City Council, the university and the Nuffield Foundation. The project aimed to research the challenges of redevelopment and then the possibilities for dealing with them, while at the same time initiating some action to transform the environment (Hilton 1967). From the beginning of the redevelopment it was recognised that it was the public sector that would be in charge, with local government taking the lead. The participation of local residents at this stage was very limited. This could be expected. No one lived in the devastated area of the floor of the valley that had been used exclusively for industrial purposes. Residents who overlooked the valley floor from the sides were only too glad to be rid of the air pollution and the desolate view. As part of the initial project, local residents were involved in environmental improvement schemes on the edge of the derelict area (Lavender 1981). Formal direct public consultation on how the valley floor should be used was not undertaken. A pressing concern of local residents was the availability of local employment. In the socio-economic climate of the area at this time it was expected that the initiatives in establishing new economic activity here would be taken by incoming private companies rather than by residents from the relatively deprived local community (Stacey 1967; Hutson and Stacey 1979). From the beginning heavy emphasis was placed on achieving economic benefits from the redevelopment. This can be seen in the way clearance of the dereliction in the Lower Swansea Valley was justified. While the despoliation of the land and the negative visual impact is always mentioned, the emphasis in the redevelopment was on making the area economically useful. It was the Lower Swansea Valley Project, that was largely based at the university and with a large university research input, which leaned towards conservation and environmental improvement. The project report was published in 1967 (Hilton 1967) and the City Council took a lead in implementing the recommendations, ably advised by the relevant officers of the administration (Bromley and Humphrys 1979). A fortuitous event also resulted in funds for waste tip clearance becoming readily available about this time. This was the Aberfan disaster in 1966 when a primary school was overwhelmed by a waste heap slide some thirty miles north east of Swansea, killing more than one hundred and twenty children. The national government readily made resources available to remove and landscape industrial spoil heaps across Wales, and these were one of the major sources of grants to clean up the Lower Swansea Valley. Without them the process would have taken very much longer. Expertise acquired locally in this kind of work was recognised throughout Europe. By 1980, most of the derelict areas had been cleared and some of them landscaped and planted with trees (Bromley and Morgan 1983). A new road system was laid out, infrastructure provided, and new commercial investment had begun to develop some of the sites.

The local authority was influenced in many of its redevelopment decisions by being dependent upon grants which were often made available where economic benefit were the criteria for the provision of the funds (Ward 1979; Howells 1979; Bromley and Morgan 1985). Even so, within the context of the economic development emphasis, the local authority could be seen to want environmental improvement. The Howells plan for the layout of the sites for commercial development on the valley floor (Howells 1979) contains strong evidence of this in the way in which sites were to be laid out in a park; wide kerbs and grass verges were to be a feature and a leisure area provided around a new lake. There was also some input from other government agencies on environmental upgrading. This was signposted by the Forestry Commission in its tree planting support, "40 ha have been acquired by the Corporation and leased to the Forestry Commission: it has been planted to produce a timber crop, to improve the recreational value of the (Kilvey) hill and to enhance its appearance" (Dinning 1971, 220). Similarly, the Welsh Development Agency leaned towards environmental enhancement for social reasons, "In a balanced society the standard of life is as important as the standard of living, and land reclamation is performing an important social function in Wales in helping to produce a cleaner, healthier and more desirable environment" (Welsh Development Agency 1984, 5). The presentation and representation of the environment was closely linked with the perception of the despoiled land as a potential economic resource, with economic development resulting in a better environment. The lack of direct public consultation on what was happening was not taken amiss by the local residents. The evidence of successful improvement, both environmental and economic, was appearing before their eyes and received general approval.

The era of privatism following the election of the first Thatcher government in Britain in 1979 reinforced the domination of economic and commercial development in the environmental changes taking place. There was a re-appraisal of the role of government both locally and nationally that resulted in a continual retraction of state involvement and an emphasis on the role of individual enterprise and responsibility. This approach essentially involved a shift in responsibility from the government and community to the individual. However, in practice there appeared to be retention of control and power by the central government with responsibility at the local level to bear the costs and deal with any shortfalls resulting from the system. Taken to its logical conclusion, it should have also required greater public participation in decision making at the local level and loosening of central direction. For the most part, this did not ensue. Presentation and representation of the environment remained with the controlling power of central government and it was these views that prevailed with little opportunity for significant local input. Environmental improvement seemed to be a low priority with the private sector seeking to minimise the costs of compliance with environmental regulations or interests (Prentice 1993). The reluctance of the British government to ensure conservation and environmental sustainability in the 1980s is illustrated in the absence of a requirement on developers to prepare an environmental impact assessment of what they proposed to do. This omission was unusual in the developed world and it was only when the European

Community issued guidance on this matter that the government moved in this direction in the early 1990s (Prentice 1993).

For the Lower Swansea Valley, an important change took place in 1981. The national Government took advantage of the progress made on the redevelopment to designate the Lower Swansea Valley as the first (and largest) Enterprise Zone in Britain (Swansea City Council 1981; Norcliffe and Hoare 1982). The sites available were soon taken up for private commercial development, almost entirely for retailing and offices, despite the intention of the establishment of the Zone to encourage the growth of light manufacturing. This reduced the extent to which the local authority could influence the planning of the subsequent development in the designated area, "The spatial arrangement of the stores in the retail park reflects its unplanned character" (Bromley and Thomas 1987, 14). The local authority did manage to retain a small but important environmental influence, in the requirement that there should be wide green verges along main roads and a verdant leisure area around the west side of the new lake in the centre of the valley. However, environmental sensitivity to the need for reduction of the use of resources and for environmental sustainability was clearly not in the minds of those who planned the layout of the larger retail units that were attracted to the new Enterprise Zone. These are arranged in such a way as to minimise their interconnectedness for pedestrian customers and the use of a vehicle to get from one site to the next. "Scant attention was given to the potential functional linkages which were likely to emerge between the stores" (Bromley and Thomas 1987, 14). What was evident was that the development did make use of a brownfield site for commercial development rather than it spreading to greenfield sites. All this demonstrated that top-down government was still dominant, with the environment still being seen essentially as an economic resource and quality of life relevance being given low priority.

The privatism approach to environmental issues apparent in Britain from 1979 to 1994 was out of keeping with the revolution in environmental attitudes and awareness that swept through Western societies after 1980 (Prentice 1993). Interest in conservation and landscape protection now came to the fore in Britain. There was a growing awareness of the limited land resources that were available and global environmental damage was beginning to make headline news (O'Riordan 1989b). By the end of the 1980s this resulted in a realisation "that the private costs to businesses do not always equate with their environmental costs, which are otherwise met by the wider community unless the government intervenes" (Prentice 1993, 285). There was widespread acceptance by now of the principle 'let the polluter pay' and by the early 1990s the concept of environmental sustainability was also gaining wider acceptance, though despite the government support for Agenda 21 (United Nations 1993) it only slowly affected British policy in the first half of the decade. Government presentations continued to stress economic growth benefits and representations of the environment focused on its use for such things as agricultural production and commercial potential for tourism. Minimum attention was given to central government responsibility for environmental sustainability and to the negative environmental impacts of economic development. This could be explained by the attitude of the party in power "that placed the burden for the protection of national environmental assets on individuals trapped in a web of economics, not of their making, and on local authorities whose judgements were conditioned by cash limits, rate ceilings and parochial perspectives" (O'Riordan 1989a, 397).

By 1994 the pattern of the redevelopment of the Lower Swansea Valley was established. Most of the sites had been occupied and the existing capital investment and the cost of any substantial alteration, made it unlikely that the pattern could be significantly altered. There were obvious shortcomings for consumers in the layout of the retail provision. This is not to criticise the enormous success and achievements in redeveloping the area but to draw attention to the lack of a contribution by direct public participation in the democratic decisions that were made.

The Coastal Zone

The post-1960 evolution of the coastal zone was different. Here it was a case of continual economic change accompanied by a growing perception of the need to improve the local environment and of the environmental impact of any local development. In the 1960s the last of the local tinplate works closed and in the 1970s the carbon black factory and the electricity generating station closed. At the same time, trade through the docks declined and there was rapid reduction in port-related employment. Large areas of railway lines previously used for the coal export trade were now surplus to need and the wagon repair operation closed. The port area, that had been publicly owned as part of the British Transport Docks Board, was sold to the private company Associated British Ports (ABP) in the 1980s as part of a government privatisation programme. The property development arm of ABP was Grosvenor Estates which was very well aware of the development potential of the large areas of surplus land within the port area. In 1965 the Ford Motor Company began components manufacture east of the main housing area and just beyond the city boundary (Humphrys 1972). In the 1980s a second aluminium wire and cable factory was built, further extending the urbanised area east of the Ford plant, but in the 1990s both the wire making factories closed.

In the 1990s two new developments impacting on the environment raised the ire of local residents. In the process of upgrading the sewage system in Swansea, a new disposal plant was built on the eastern edge of the port area. Less welcome was the location of an incinerator in the area in 2002. This was built immediately east of the Swansea City boundary to process domestic waste from other local authority areas on the east side of Swansea Bay. After 2000, major change began to take place in the port area. Public funds were used to build two techniums along the left bank of the River Tawe to house high tech start-up companies and commercial funds were invested in an office block built nearby. In 2003 construction began of a new marina housing area within the port area alongside the innermost dock, mirroring an earlier successful marina/housing development on the west side of the River Tawe in the 1980s. The left bank development will eventually comprise more than fifteen hundred housing units in a mixture of housing and apartments along the waterside, together with small retail units and restaurants (Welsh Development Agency 2004).

These developments began to enhance the landscape of the coastal zone and to shift the perception of the area from one of low-grade urban development to one of new urban growth and attractiveness. There was a 'halo' effect on property prices in the existing housing areas. Average house prices in the coastal zone trebled between 2000 and 2004. This had a social impact. One of the local elected councillors for the Ward commented that, "The changed housing situation caused changes in social attitudes and difficulties for children of local residents trying to buy their own homes for the first time" (Alan Richards, Elected Member for St Thomas Ward on Swansea City Council, personal communication). Previously rented properties were sold off as the owners realised their assets and local residents seeking to buy a house for the first time were priced out of the market. Existing and incoming house owners were now much more conscious of the value of their properties as capital assets and more sensitive to environmental changes which might have an impact on this.

Most of the pre-2002 development in the coastal zone took place within or in close proximity to the housing areas. Here the changing social attitudes began to be evident with the appearance of local community environment action groups. One factor which has long played a part in environmental politics affecting Swansea East is the local geography. Swansea West and Swansea East differ markedly in their social, economic and environmental characteristics. The prevailing wind blows from the west and it is in the west that the more prosperous suburbs are located. These contrasted with the polluted and scarred industrial landscapes that had been prominent in Swansea East. There is a natural questioning of those living in the latter as to why they have to put up with all the rubbish and residents are quick to point out that the planners almost invariably live in the western suburbs of the city. By 1970, local societal changes were allowing freer expression of objections to negative elements of the environment. In earlier periods when jobs were scarce, people who lived locally found work in local industries. Even if these had detrimental effects on the environment, people were reluctant to protest because of the possible threat to the jobs of their neighbours. In the 1970s, far fewer jobs remained locally. Rising car ownership meant that local residents more often travelled elsewhere to work and local factories employed more people from elsewhere.

Evidence of the changing social attitudes to the environment and to the way decisions were made, was seen in the case of the carbon black factory. Just before Donald Anderson was elected as Member of Parliament for Swansea East in 1974, there were demonstrations by women living near the works against the pollution emanating from the carbon black factory. They objected to the oily smuts that rained down on the clothes they hung out to dry and on their cars. The company undertook a hearts and minds campaign but made little headway. The leadership of the political campaign against the owners was very much indigenous and outside the formal political structures. The newly elected MP had just returned from having previously served as a local councillor in London where housing protesters in the ward he represented were led by middle-class progressive outsiders. Undoubtedly a factor in the difference was the fact that in the London ward only 5 per cent of the heads of households were born in the borough, whereas in Swansea East the majority were born locally and formed a vibrant community. The carbon black factory eventually closed and, partly influenced by the actions the local community had taken, the site was redeveloped as playing fields for the local community (Donald Anderson, Member of Parliament for Swansea East, personal communication).

In the 1990s it became apparent that there was a need for a new sewage processing plant for Swansea. A site for a new works was identified in Swansea East near the eastern end of the docks. The public debate about where to site the works aroused strong passions in Swansea East. Residents there felt that they were once again being disadvantaged compared to Swansea West. Eventually, with reassurances about landscaping the site, the works was located in Swansea East. When completed, the landscaping was to such a very high standard that it enhanced the eastern approach to the city. In this case, if the protesters had been successful, the area would have missed out on an environmental improvement. A further landscape enhancement occurred in 2003 on a site adjacent to the sewage works when the series of ten-metre-high oil storage tanks built in the 1920s were demolished. No objections to the clearance were raised by local residents though the site has yet to be landscaped. Both these changes had a positive impact on the landscape of the coastal strip.

Less welcome was the construction of a waste incinerator in 2002, very close to the eastern boundary of Swansea but just over the border in the County of Neath and Port Talbot. There were very strong protests from those living nearby on both sides of the boundary (Stop the Incinerator Campaign 2002). They used the Internet to develop considerable expertise in the science of waste incineration and emissions. Unfortunately, there were insufficient grounds for it to be rejected on planning grounds and it was built. However, the objectors continued their fight and using their newly developed expertise have pressured the authorities to monitor closely the level of polluting emissions, causing the works to close for modification on occasions.

A very different controversy affecting the coastal zone emerged during the late 1990s. For many years sandbanks offshore have been dredged as a source of sand for construction purposes. Since 1990 sand has been disappearing from all along the coast and particularly along the south Gower coast in Swansea West. Objectors have opposed the renewal of licences for further dredging but without success as the Welsh Assembly has now approved dredging until 2008. The difficulty here is that there is insufficient understanding of the complex dynamics of water and sand movement offshore to know what the long-term effects of the sand removal will be. The objectors seek the application of a precautionary principle with the cessation of dredging until more knowledge is available.

Two further environmental issues emerged after 2000. On-shore there is continuing objection to the erection of phone masts by the mobile phone companies. Most people find these unsightly and there are fears about health hazards from the signals emanating from these masts, especially for young children. Despite the lack of evidence of danger, local councils are being much more careful when dealing with applications for their siting. Off-shore developments have been proposed which would utilise the potential for electricity generation in Swansea Bay. An application submitted for the erection of a major wind farm to the south east of the Bay was the subject of a planning enquiry in early 2004, the outcome of which had not been announced at the time of writing. If approved, the windmills would be up to 100 metres high and very visible from Swansea East. Meanwhile, investigations have begun for the location of lagoons in the Bay much closer to Swansea East. These will take advantage of the second highest tidal range in the world, here averaging eleven metres and exceptionally being over fourteen metres, to generate electricity (*South Wales Evening Post* 2004). Objectors see their best chance of success being related to the navigation hazards these might create though they are also concerned about the visual impact of such a development.

Conclusion

The environmental experience of Swansea East since 1960 shows how societal attitudes towards, and assumptions about, environmental matters have undergone substantial shifts. It also shows that the political response to reflect the changes has been very much slower and only now is showing signs of catching up. The major shift has been in the local awareness of the importance of the environment and of changes in it both generally and as an important local quality of life factor. Along with this shift has come much more widespread understanding of environmental matters among local residents, aided by improving standards of education and an increasing proportion of the population with formal qualifications. These changes were evident much earlier in the more affluent areas of the conurbation than in Swansea East which was and still is a largely working class society. The changes also influenced attitudes to and understanding of the way democratic government worked, though there was continued acceptance that it was elected representatives who took the decisions, with the influence of the electorate primarily expressed through the ballot box. Nevertheless, the actions taken to protest about the carbon black plant emissions in the 1970s were an indication of how perceptions and attitudes had changed even then. It was a presentation and representation of the environment which differed from that of the company involved and provoked further action by the local authority. A later major contribution to how people could take advantage of their greater environmental interest and understanding, was the rapid increase in the availability of, and access to, information via the Internet after 1990. This meant that the scientific knowledge on which professional experts depended was readily accessible to all and its built-in values and limitations readily exposed. In a sense it was democratised. The asymmetry of power was being eroded. This asymmetry exists because those with sufficient resources available, such as large industries or developers, have always been able to influence decisions by bringing to bear professional advice accepted as authoritative, while opponents lacked the resources to respond adequately. Similarly, governments at different levels also use their professionals to justify a decision and argue against those opposing it. The domination in the presentation and representation of the environment of the value judgments of those with power was now exposed and could be challenged. A better educated public with ready access to knowledge at least reduces the unequal advantage. This was particularly well illustrated in Swansea East by the well organised group opposing the building of the incinerator and then ensuring that its emissions were monitored and that the strict limits on the emissions were observed. In such situations there is the possibility that fringe groups with other motives may take advantage of local action to promote their own agenda. The best defence against this is the improving levels of education of the public.

The implications for democratic government of a more environmentally aware, educated electorate with ready access to high quality relevant information are now becoming apparent (Stoker 2000). There is a need for better ways to be provided for their presentation and representation of the environment to be heard and taken into account when environmental decisions are being taken. The presentation and representation only by their elected representatives no longer seems sufficient. In the Swansea East example, it is likely that if this had been possible earlier it may not have confirmed the way the pre-1980 development of the Lower Swansea Valley took place. It almost certainly would have impacted on the post-1980 retailing layout there. For the coastal zone it would have been and is more important. There is no doubt that the local elected representatives have striven to undertake properly their responsibilities but structures are not in place to allow more direct public participation in the democratic process. What is promising is the attitude of the British Government and the Welsh Assembly Government as revealed in their publications and actions. These show a much greater sensitivity to the heightened awareness and understanding of the environment by the electorate and of the need to increase public participation (See, for example, Office of the Deputy Prime Minister 2002 and 2003; Wales Assembly Government 2003; Wales Assembly Government 2004). The way these present and represent the environment and see the need for other views to be taken into account by government is a very encouraging beginning. They indicate a promising future for a more modern democratic approach to dealing with local environmental matters. Nevertheless, there is still some way to go in improving both the presentation and representation of the environment in the political arena and the role of public participation within the decision making process before a satisfactory system of good governance is achieved.

References

Barton, B. (2002). Underlying concepts and theoretical issues in public participation in resources development. In D. N. Zillman, A R. Lucas, and G. Pring (Eds.) *Human Rights in Natural Resource Development: Sustainable development of mining and energy resources.* Oxford: Oxford University Press.

Bromley, R. D. F. and Humphrys, G. (Eds.) (1979). *Dealing with Dereliction: The redevelopment of the Lower Swansea Valley.* Swansea: University College of Swansea.

Bromley, R. D. F. and Morgan, R. H.(1983). *Change and Industrial Development in the Lower Swansea Valley.* Swansea: Department of Geography, University College of Swansea.

Bromley, R. D. F. and Morgan, R. H. (1985). The effects of Enterprise Zone Policy: Evidence from Swansea, *Regional Studies*, 19, 403–413.

Bromley R. D. F. and Thomas, C. (1987). *Retail Parks, Enterprise Zone Policy and Retail Planning: A case study of the Swansea Enterprise Zone Retail Park.* Working Paper Number 1, Social-Economic Research and Environment. Swansea: Department of Geography, University College of Swansea.

Dinning M. (1971). Forestry. In W. G. V. Balchin (Ed.) *Swansea and its Region.* Swansea: University College of Swansea.

Dryzek, J. S. and Schlosberg, D. (Eds.) (1998). *Debating the Earth: The environmental politics reader.* Oxford: Oxford University Press.

Funtowicz, S. O. and Ravetz, J. R. (1993). Science for the post-normal age, *Futures*, 25, 739–755.

Held, D. (1996). *Models of Democracy.* (2nd Ed.) Cambridge: Polity Press.

Herbert, D. T. (1971). The twentieth century. In W. G. V. Balchin (Ed.) *Swansea and its Region.* Swansea: University College of Swansea.

Hilton, K. J. (Ed.) (1967). *The Lower Swansea Valley Project.* London: Longmans.

Howells, M. (1979). Planning of the valley since 1974: Physical plans with a social meaning. In R. D. F. Bromley and G. Humphrys (Eds.) *Dealing with Dereliction: The redevelopment of the Lower Swansea Valley.* Swansea: University College of Swansea.

Hughes, S. R. (2000). *Copperopolis: Landscapes of the early industrial period in Swansea.* London: Royal Commission on the Ancient and Historical Monuments of Wales.

Hulme, M. and Turnpenny, J. (2004). Understanding and managing climate change: the UK experience, *The Geographical Journal*, 170, 103–115.

Humphrys, G. (1972). *Industrial Britain: Industrial South Wales.* Newton Abbot: David and Charles.

Humphrys, G. (1990). Twentieth century change. In G. Williams (Ed.) *Swansea: An illustrated history.* Sketty, Swansea: Christopher Davies.

Hutson, S. and Stacey, M. (1979). Living in the valley. In R. D. F. Bromley and G. Humphrys (Eds.) *Dealing with Dereliction: The redevelopment of the Lower Swansea Valley.* Swansea: University College of Swansea.

Lavender, S. J. (1981). *New land for Old: The environmental renaissance of the Lower Swansea Valley.* Bristol: Adam Hilger.

Marsh, D. and Stoker, G. (1995). *Theory and Methods in Political Science.* Basingstoke: MacMillan.

Muller, A. (2003) A flower in full blossom? Ecological economics at the crossroads between normal and post-normal science, *Ecological Economics*, 45, 19–27.

Norcliffe, G. B. and Hoare, A. G. (1982). Enterprise Zone policy for the inner city: A review and preliminary assessment, *Area*, 14, 265 –274.

Office of the Deputy Prime Minister (2002). *Community Involvement: The roots of renaissance?, Urban Research Summary Number 5.* www.odpm.gov.uk/urban/research (accessed July 2004).

Office of the Deputy Prime Minister (2003). *Searching for Solid Foundations: Community involvement in urban policy. Urban Research Summary Number 10.* www.odpm.gov.uk/urban/research (accessed July 2004).

O'Riordan T. (1989a). The challenge of environmentalism. In R. Peet and N. Thrift (Eds.) *New Models in Geography, Vol. 1.* London: Unwin Hyman.

O'Riordan, T. (1989b). Contemporary environmentalism. In D. Gregory and R. Walford (Eds.) *Horizons in Human Geography.* London: Macmillan.

Paehlke, R. and Torgeson, D. (Eds.) (1990). *Managing Leviathan: Environmental politics and the administrative State*, Peterborough, Ont.: Broadview Press.

Prentice, R. (1993). *Change and Policy in Wales: Wales in the era of privatism.* Llandysul: Gomer.

Ravetz, J. R. (1999). Post normal science, *Futures*, 31 (Special Issue.), 641–757.

South Wales Evening Post (2004). *Strong Tidal Pull.* Swansea: *South Wales Evening Post*, 4 November.

Stacey, M. (1967). The people, their environment and their houses. In K. J. Hilton (Ed.) *The Lower Swansea Valley Project.* London: Longmans.

Stoker, G. (Ed.) (2000). *The New Politics of British Local Governance.* London: Macmillan.

Stop the Incinerator Campaign a (2002), (STIC) www.stic.org (accessed September 2004).

Swansea City Council (1981). *Swansea Enterprise Park – Britain's First Enterprise Zone.* Swansea: Swansea City Council.

United Nations (1993). *Agenda 21: A Programme of Action for Sustainable Development: Rio Declaration on Environment and Development: The Final Text of Agreements Negotiated at the United*

Nations Conference on Education and Development (UNCED), 3–14 June, Rio de Janeiro. New York United Nations agreement signed at the 1992 Rio Summit.

Wales Assembly Government (2004). *People, Places, Futures: The Wales spatial plan*. Cardiff: Welsh Assembly Government.

Wales Assembly Government (2003). *Starting to Live Differently: Consultation on the review of the Sustainable Development Scheme*. Cardiff: Welsh Assembly Government

Ward, W. J. (1979). Planning of the valley before 1974. In R. D. F. Bromley and G. Humphrys (Eds.) *Dealing with Dereliction: The redevelopment of the Lower Swansea Valley*. Swansea: University College of Swansea.

Welsh Development Agency (1984). *The Changing Face of Wales*. Pontypridd: Welsh Development Agency.

Welsh Development Agency (2005). SA1 Development. www.wda.co.uk (accessed January 2005).

Williams, D. T. (1940). *The Economic Development of Swansea and of the Swansea District to 1921*. Cardiff: University of Wales Press Board.

7. SUSTAINING LOCAL RIVERINE ENVIRONMENTS: THE RIVER VALLEYS COMMITTEE IN CALGARY, ALBERTA, CANADA

DIANNE DRAPER

Wild rivers, sparkling lakes and immense glaciers are an integral part of Canadian life and identity. Abundant, high quality water resources have sustained Canada's quality of life for centuries; our lakes and rivers have provided a template for the settlement patterns of the country and have sustained the many economic and recreational activities of Canadians. Yet, in only one generation, concerns about the availability, supply and quality of Canada's water resources have challenged long-held perceptions and required new ways of visualising, representing, understanding and managing water resources. These concerns are, perhaps, most evident in western Canada's prairie provinces where population growth, drought conditions, climate variability and change, agricultural and industrial development and urbanisation have placed increasing pressure on water resources. Political debate about these water concerns frequently has pitted environmental protection against economic prosperity. However, if water (and other) resources are to be valued as natural capital assets, they must be recognised as such and embedded within public policy decisions. Even though western Canadians place a high priority on environmental protection, ensuring that public policy frameworks integrate stewardship of natural capital with economic growth remains a political challenge.

Natural capital is a relatively new term that is entering public policy discussions about the need to invest in protection of water, land and other environmental resources that sustain our way of life and support economic activities. Often, reconciling resource development with preservation of natural capital depends on effective community consultation and involvement. Since 1990, Calgary's River Valleys Committee (RVC) has been working to protect and enhance the city's riverine environments and to ensure that City Administration is encouraged and exhorted to develop and implement effective mechanisms to protect and sustain water resources as the city grows. This chapter focuses on the establishment and activities of the RVC, how its members are helping to present water resources as natural capital assets and how this representation has influenced public policy regarding water resources as natural capital. The chapter first describes the nature of, and some of the major pressures

G. Humphrys and M. Williams (eds.), Presenting and Representing Environments, 105-119.

affecting, water resources in southern Alberta and the City of Calgary and considers the importance of water resources as natural capital. The natural capital concept is explored in more detail and is followed by a brief history of the River Valleys Committee and its activities relative to water resources within Calgary. Throughout, an effort is made to identify how the local environment is presented and represented, and the implications this has for water resources in the local area and beyond.

Water Resources and Human Activities in Southern Alberta

Southern Alberta stretches from Red Deer south to the Montana border (Figure 7.1). Except for a narrow belt of aspen parkland in the north, and mountain and foothill landscapes to the west, the southern Alberta prairie landscape consists predominantly of grassland ecosystems and is characterised primarily as a semi-arid area with abundant sunshine, strong dry winds, and an average annual precipitation of 300–450 millimetres. This precipitation originates in the Pacific Ocean and falls as snow or rain in the Rocky Mountains and contiguous foothills. Spring snowmelt from these high elevation headwaters runs off the eastern slopes of the Rockies and provides the bulk of Alberta's surface water flow (over 97 per cent of the water Albertans use comes from surface water sources; 2.5 per cent is from groundwater). Over 85 per cent of the water in Alberta's rivers flows north into the Arctic Ocean via the Athabasca, Peace and Slave River watersheds, while less than 15 per cent flows east into Saskatchewan via the Churchill and South Saskatchewan River basins. Water in the Milk River basin flows south to the Gulf of Mexico (Corbett and Lalonde 2004; Harras 2003).

Constituting only 2 per cent of southern Alberta's land, riparian areas in the mountains, foothills and grasslands sustain 80 per cent of southern Alberta's wildlife, including nesting songbirds, deer and other mammals, several trout species and other fish. The warm, dry southeastern portion of the province supports pronghorn antelope, badgers, prairie rattlesnakes and prairie falcons. Overall, the "native prairie wildlife community is distinctive in North America because of its historic assemblage of large mammals. The prairie pothole region is also one of the most productive waterfowl breeding habitats on the continent" (Corbett and Lalonde 2004, 3). Southern Alberta's riparian forests also are unique, dominated by poplar trees and large stands of plains cottonwood along the prairie reaches of the South Saskatchewan River system. Riparian forests in the prairies function to store carbon, recharge aquifers, trap sediment and stabilise banks, important environmental (natural capital) services.

The Bow, Oldman and Red Deer Rivers feed the South Saskatchewan River, which drains 18 per cent (120,000 square kilometres) of the province. Providing just 7 per cent (about 9,000,000 cubic decametres) of Alberta's total annual surface flow, the South Saskatchewan River basin supports about half the province's population. While southern Alberta occupies less than 20 per cent of the total area of the province, it is home to more than one and a half million residents, 81 per cent of whom live in urban centres along the major rivers and two main highways. At close to one million resi-

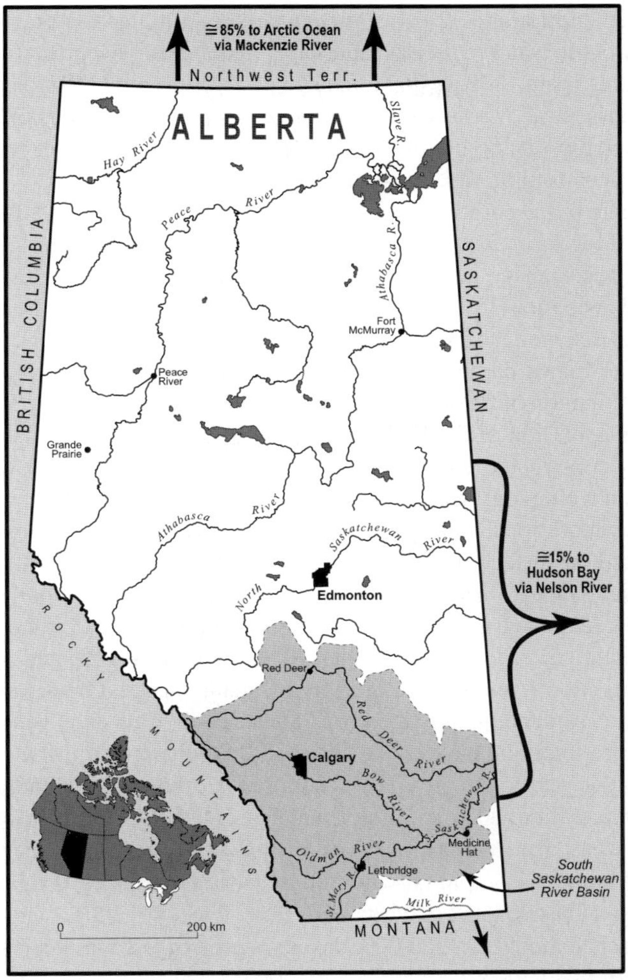

Figure 7.1 *Southern Alberta and the South Saskatchewan River Basin.*

dents, Calgary (Canada's fifth largest city) dominates the region, but all of the cities, including Lethbridge, Red Deer, and Medicine Hat, have experienced considerable population growth during the past decade. About half of the province's economic activity is generated in southern Alberta, principally in the petroleum, agriculture, manufacturing and service sectors (Corbett and Lalonde 2004). In the larger cities, knowledge-based and value-added processing industries also are major drivers of the provincial economy.

Because the arid southeast area of the province experiences more evaporation than precipitation (the average net moisture deficit is 400 millimetres), irrigation works have been vital to supply water to agricultural, industrial and municipal users in the region (Corbett and Lalonde 2004). Pioneer farmers initiated the practice of water diversion to provide water to the dry landscape, undertaking Alberta's first irrigation project near Calgary in 1879. In 1894, faced with multiplying diversions, the federal government passed the *Northwest Irrigation Act*, simultaneously establishing government ownership of and control over water resources, and allocation of water licenses on a first-in-time, first-in-right basis.

Subsequent to failed corporate investments in irrigation canals and diversions, the 1915 Alberta *Irrigation Districts Act* permitted landowners to form local cooperatives to build and manage large irrigation projects. Unique to Alberta, the thirteen farmer-owned irrigation districts eventually covered over half a million hectares of irrigated land. However, the irrigation districts also faced financial difficulties as they constructed the major capital works for water storage and delivery. Ultimately, through a combination of federal and provincial funds, made available from the time of the Great Depression of the 1930s until the early 1990s when a large dam and reservoir was constructed on the Oldman River, irrigated acreage continued to increase. Use of technology such as centre pivot sprinklers, and infrastructure repairs that reduced evaporation and seepage losses, contributed to this expansion.

Today, based on surface water supplies, 4 per cent of Alberta's total arable land reliably produces about 20 per cent of the provincial agricultural output, even though these lands are exposed to considerable periods of drought. Southern Alberta's irrigation agriculture represents more than 70 per cent of the irrigated farm acreage in Canada: irrigation systems have anchored Canada's largest beef cattle industry, other livestock, produce and forage crop industries as well as expansion in food processing operations, and have added about 35,000 jobs and contributed about one billion dollars per year to the provincial economy (Corbett and Lalonde 2004).

Larger municipalities also are major water users: Calgary, for instance, began withdrawing water from the Bow River in the 1890s and continues to rely on river waters to support its rapid growth. For the past two decades, Calgary's population growth rate has led the nation; between 1996 and 2001, when Calgary's population reached more than 950,000, the city's growth rate was nearly 16 per cent (Worbets and Berdahl 2003; City of Calgary 2003a). The unprecedented growth rate (1999 to 2003) equated to 56 new people arriving in Calgary daily; every day these new residents required 21 additional dwelling units and put 36 more cars on the roads (City of Calgary 2003b). Despite this population growth, per capita consumption of water has declined from close to 800 litres per person per day in 1979 to an average of 516 litres per person per day in 2003. Continued use of both regulatory and technological advances will assist Calgary's Waterworks division to achieve its goal of accommodating population growth using the same amount of water in thirty years time as is used today: the average use will need to be 339 litres per person per day, about one-third less than is consumed now (City of Calgary, Waterworks 2004).

Physical expansion of the city has resulted in an increasingly larger 'footprint' including the consumption of between 80 and 90 per cent of the prairie wetlands that once dotted the Calgary landscape (Parks Foundation, Calgary 2004). Accommodating urban growth in Calgary by filling in and paving over wetlands (sloughs, marshes, seasonal ponds) negatively affects such natural capital functions as water purification processes and recharge of groundwater resources. These are among the concerns that Calgary's River Valleys Committee attempts to bring to the attention of City Council and administrators and the public.

During recent decades, many of the river courses flowing through southern Alberta's cities have been transformed from industrial backyards to aesthetically pleasing public parkway systems, providing numerous recreational activity opportunities; for instance, the Bow River immediately downstream of Calgary, and the Oldman River and its tributaries, support world-class trout fisheries. Yet, members of the public increasingly have become concerned about water quality issues in the South Saskatchewan River Basin, including drinking water quality and ability of the region's rivers to assimilate municipal effluents. Treatment of municipal wastewater in Alberta requires, at minimum, use of secondary treatment processes to ensure 90 per cent removal of organics and solids, as well as disinfection and removal of nutrients to reduce effluent loadings (Corbett and Lalonde 2004). While all cities in southern Alberta have invested in improved sewage treatment, Calgary's tertiary treatment system has been rated the best of all major cities in Canada. Other point and non-point pollution sources, and alterations to natural flow, however, remain active concerns for the River Valleys Committee (and others).

Governance and Allocation of Water Resources in Alberta

Under the *Water Act*, the Alberta government owns the rights to all water within its borders, the federal government having ceded control of water resources to the province in 1931. Through legislation, Alberta regulates activities with potential effects on rivers, lakes and groundwater, while the federal government retains jurisdiction for interprovincial and international water agreements, navigable waters, fish habitat protection and rivers in national parks, where the Bow and Red Deer Rivers have their headwaters (Alberta Environment, Bow River Basin Council 2002). While municipalities have some ability to regulate their water use, the provincial government takes the lead role in response to the increasing pressures being placed on southern Alberta's water resources.

The provincial government allows water diversion and use by license holders; the first-in-time, first-in-right basis of the licensing system means that in times of water shortages, the oldest licenses have highest priority, regardless of the purpose for which the water was allocated. With some licenses dating back to 1894, currently there are about 20,000 licenses and registrations in the South Saskatchewan River Basin; these licenses account for 61 per cent of all the water allocated in Alberta (Corbett and Lalonde 2004). In the South Saskatchewan River Basin, 75 per cent of the allocation volume (more than 3.8 million decametres of water) supports irrigation.

Not all the water that flows in southern Alberta's rivers is available to supply local use or instream needs. A 1969 interprovincial agreement requires Alberta to pass approximately 50 per cent of the South Saskatchewan River's natural flow to Saskatchewan (the water must meet certain quality guidelines also). On average, Alberta has passed approximately 75 per cent of the flow to the east but, in 2001 (one of the driest years on record) just over 50 per cent flowed to Saskatchewan (Corbett and Lalonde 2004). The Boundary Waters Treaty of 1909 established flow requirements for the Milk and St. Mary's Rivers that cross the Alberta-Montana border.

Once the international and interprovincial agreements are satisfied, the water remaining in southern Alberta's river basins is available for instream needs, including hydroelectric power generation, transportation, fisheries and recreation, and for allocation to withdrawal uses such as irrigation, oil and gas well injection to enhance recovery rates and thermal power generation. High allocation rates on some of southern Alberta's rivers threaten the future availability of sufficient, good quality water to meet instream needs. In the Bow and Oldman River systems, for instance, instream flow needs for water quality, fish habitat, riparian vegetation and channel structure have been determined to require flows much greater than existing flows. Since restoring required flows would be impossible under the present allocation system, some professionals believe that the long-term health of the aquatic environment in some reaches of these rivers is declining (Golder Associates Ltd. 2003). Conversely, in some river reaches such as the Bow River south of Calgary, enhanced wastewater treatment has improved aquatic habitat conditions and resulted in growth in trout populations. The present demand for water in Alberta is met principally through withdrawals from the South Saskatchewan River system. Although the historic average natural flow in the South Saskatchewan River basin has not changed much, highly variable flows on the tributaries of the river system complicate the management of water resources. Low precipitation in recent years, coupled with the loss of reliability of the glacial water supply ('fossil water') from the Rockies, due to climate change (Nikiforuk 2004), means that historical flows cannot be assured in the future. If climate change reduces water flows in Alberta's rivers, water shortages not only will affect agricultural and industrial activities on the prairies, but also urban centres such as Calgary will receive less water.

The preceding overview of the nature, governance and allocation of southern Alberta's water resources, as well as human activities in the South Saskatchewan River basin, was intended to demonstrate that there are significant challenges in managing southern Alberta's water resources, both now and in the future. Recognising that a comprehensive, integrated approach to natural resource (natural capital) and environmental management is required, it is equally important that shared stewardship – with all affected and interested parties involved in planning, decision making, implementing and monitoring progress toward sustainability – be achieved. The example of commitment to natural capital stewardship provided by the efforts of members of Calgary's River Valleys Committee highlights the role that citizens may play in contributing to public policy. The following section describes the natural capital concept and its importance in public policy regarding water resources.

The Natural Capital Concept

In our globalising world, it is recognised clearly that a healthy environment and the high quality of life that it offers represent part of a nation's or region's competitive advantage in attracting and retaining skilled workers (McGuinty 2002). Natural capital (resources which provide raw materials for production of manufactured goods, as well as water, land and living ecosystems that contribute to quality of life and support economic activity) is one of four kinds of assets employed by nations and regions in the development of prosperous economies and an enhanced quality of life. Produced capital (including the financing, equipment and infrastructure that produce goods and services for businesses and consumers) and the ongoing investments required to counteract the deterioration of these assets that occurs over time have been identified and incorporated in public policy frameworks for a long time. Human capital (the skills and health status that individuals bring to an economy) and social capital (community attributes such as trust and social cohesion) have been part of public policy frameworks to a lesser extent and for a shorter period of time than produced capital.

Ideally, since all forms of capital are necessary for both economic prosperity and quality of life, public policies should attempt to protect nature's assets (natural capital) as effectively as they protect produced capital assets. However, there has been less political recognition of the importance of natural capital compared to other forms of capital assets and Canadian public policy frameworks, for the most part, have not ensured a balance among natural, produced, social and human capital assets (England 1998; Worbets and Berdahl 2003). Often, as England (1998) notes, assessments of capital stock fail to include the contributions that nature's assets provide, suggesting not only that the concept and measurement of capital needs to be broadened to include natural capital assets but also that western Canadian provinces "have become successful at the expense of our land and water resources" (Worbets and Berdahl 2003, 13). For instance, the 2003 Alberta budget documents included many measures of economic and social performance, but "not a single measure of natural capital" (Alberta Finance 2003; Worbets and Berdahl 2003, 3). Furthermore, since natural capital assets are subject to deterioration through resource extraction, excessive growth, waste and landscape modification, the long-term interests of the environment need to receive protection within political systems, public policy frameworks and management regimes. Investment in the protection and preservation of natural capital assets is essential to ensure environmental sustainability and prosperity.

The natural capital concept "provides a different way of thinking about the environment that focuses on its importance to both our economic prosperity and our overall quality of life" (Worbets 2005, 2). By helping to "close the counterproductive gap that exists between environmental interests on the one hand and business interests on the other" the natural capital concept attempts to address the misperception that investment in the environment (natural capital) does not make economic sense and that the environment inevitably is harmed as the economy grows (Worbets 2005, 2). Fundamentally, natural capital is a primary economic input and needs to be at the core of policy and planning processes, particularly those pertaining to water and land resources.

In this respect, and particularly with respect to water resources, all urban, working and wild landscapes need protection. Urban landscapes, typically thought of in terms of their skyscrapers, concrete and traffic congestion, also may be home to parks, viewscapes, rivers and lakes, forests and wildlife, all of which contribute to the quality of urban life and all of which reflect natural capital assets. Calgary, for example, contains more than 11,000 hectares of parks, open and green spaces, and over 550 kilometres of pathways connecting these parks and green spaces – that 90 per cent of residents use regularly (City of Calgary 2002). Working landscapes – the vast ranches; wheat, barley and canola fields that characterise much of southern Alberta – while they may be seen as intrusions on the natural landscape, also are inherent to the western heritage, beauty and identity of the region as well as its economic prosperity.

Perhaps unparalleled in the world, Canada's living, wild landscapes are an intrinsic part of the Canadian identity. Including national and provincial parks, 'wilderness' areas are valued for their scenic beauty, for the wildlife and habitats they contain, as locations for recreational activity that contribute to physical wellbeing and quality of life for residents and visitors, and for the contributions that tourism operations in wild landscapes make to the economy. Of particular significance, given that the headwaters of the South Saskatchewan River are located in the Rockies and foothills, is that preservation of wild landscapes (for example mountains as 'water towers') helps secure the availability of future water resources for southern Alberta users.

Recent opinion surveys of western Canadians have revealed that protecting the environment continues to be a priority and that environmental protection (including use of water conservation measures such as full-cost pricing) is not a barrier to strong economic performance (Berdahl 2003). Yet, it also is clear that neither all citizens nor all politicians or bureaucrats understand the necessity to have the long-term protection and management of water resources upheld within a natural capital public policy framework. Key elements in a natural capital framework include: recognition of the value of water and a commitment to protect and sustain it; empirical measurements of the health of water and other natural capital assets as a measure of progress toward sustainability; a commitment to protect, manage and invest in building natural capital; development of a vision and strategy for sustaining water resources; and coordinated policies and practices for water and land uses (Worbets and Berdahl 2003). Thus, the public educational roles of the Calgary River Valleys Committee and its contributions to water and land use policy decisions in the City reflect important components of a natural capital framework. The following section discusses the River Valleys Committee, its establishment, water and land use stewardship activities, and efforts to have local water resources represented and understood as natural capital assets.

The Calgary River Valleys Committee

Calgary's River Valleys Committee is a volunteer-driven organisation charged with understanding, protecting, nurturing and championing the City's river system for the continued benefit of all Calgarians, downstream residents and users, and visitors to

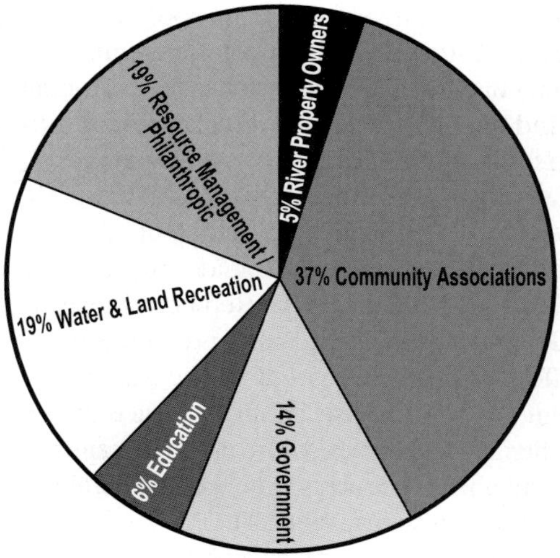

Figure 7.2 *Membership groups of the Calgary River Valleys Committee.*

the region. Today, with a fifteen member Executive Committee, twelve member Advisory Board, three sub-committees and over sixty diverse member groups (Figure 7.2), the RVC's core activities include:

a. actions undertaken to educate and enhance awareness, and to promote good stewardship and adoption of best practices among the public, community groups, planners, developers and all orders of government, regarding the rivers, river valleys and their watershed;
b. projects that protect, preserve, acquire, reclaim or establish key river basin assets, including the diversion of inappropriate uses to other suitable locations;
c. policy, planning and development review that includes participation in the development of appropriate policies and plans, evaluation of development proposals, review of land and water uses, land and water management practices, and that promotes interaction and integration among all orders of government and their departments and agencies;
d. promoting and supporting land acquisition and stimulating donations of river valley and watershed lands (River Valleys Committee 2003, 3).

The idea for the RVC originated in July 1989, when then Alderman Al Duerr identified the need for a body to provide "direction and focus to the care and management of Calgary's river valleys" (Davis 2000, 2). Citing multiple City Council approved documents that related to the city's rivers and valleys, including the 1984 *Calgary River Valleys Plan*, Duerr noted the need to "champion the cause of protecting and enhancing the Calgary River Valleys as a resource and a legacy for the future" and asked the City Administration to develop a programme so that the public could participate in developing and maintaining the city's river valley system (Davis 2000, 2).

In September 1989, a City report recommended that representatives from all interest groups should meet to discuss options for forming a 'River Valley Action Group' and to consider a strategy to develop the group's programme and mandate. In April 1990, the Parks Foundation, Calgary struck a special 'River Valleys Committee' to form a partnership with all stakeholder volunteer groups and government bodies (as resources). (The Parks Foundation, Calgary, is a non-profit, non-governmental organisation founded in 1985 to enhance the quality of life for all Calgarians through the development of parks, river valleys and amateur sports).

In September 1990, a City Supplemental Report identified the mission and terms of reference of the RVC. The RVC's original mission was to "enhance the profile of the city's river valleys for the purpose of protecting and improving the overall riverine environment" and to "advise on and advance the development, use and conservation of the city's watercourses, and to carry out capital project enhancements for river bank parks and open space areas from funding obtained from various private and public sources" (City of Calgary 1990, 1).

The River Valleys Committee was placed within the organisational structure of the Parks Foundation, Calgary, and the original agreement identified a formal reporting process for the RVC (through the Foundation's Board of Governors) directly to City Council. The policy direction given to the RVC was to use the 1984 *Calgary River Valleys Plan* (City of Calgary 1984) as a starting place for its specific role and mandate, but it was clear that the RVC could extend its interest beyond the focus of the river valley system where an adjacent development could be shown to impact on the system. In addition, where project proposals were reviewed, the RVC could comment directly to the Planning department and to City Council. The RVC's direct access to City Council and Planning department, the breadth and policy focus of the RVC mandate, and the ongoing financial support of the Parks Foundation, Calgary, have been critical elements in the Committee's success to date.

From its outset, the RVC operated as a catalyst in joint partnership with communities, individuals and other volunteer groups to share ideas and to take action on water and land issues affecting the river courses in Calgary. Indeed, the RVC has been successful in attracting a wide range of members and subsequent project partners; the value of these partnerships for the City is that they facilitate effective and efficient access to Calgary's communities for the City (and provincial) Administrations. Similarly, for Calgary's communities and citizens, the RVC provides ready access to local and regional governments. Without the RVC, the City would need to reach each of the 40 river communities and many other stakeholders individually, increasing time and costs of communication.

As stated previously, natural capital needs to be at the core of policy and planning processes, and the RVC (viewed as an organisation and as reflective of the qualities of its active members) brings not only the concepts of natural capital to bear on policy issues relating to Calgary's water and land resources, but also draws on human capital (professional skills) and social capital (community partnerships) elements. That is, the unique position the RVC occupies as a direct partner with the City of Calgary, the province (Alberta Environment) and the federal government (Department of

Fisheries and Oceans) provides opportunities to caring citizens for credible and meaningful involvement in the affairs of their community. This participatory opportunity results in a high degree of credibility for the RVC and, as a result, "some of the brightest and most dedicated professionals in business, education and government in Calgary" are attracted to the RVC, in turn enhancing the credibility of the RVC's contributions to public policy processes (River Valleys Committee 2003, Attachment 2, 4). Many of these individuals are highly knowledgeable about water resources, about the state of water resources in southern Alberta, about planning processes in Calgary and the province, and use their existing professional and political networks to help achieve the RVC's public policy objectives.

While natural capital terminology *per se* was not employed in the RVC's mandate, mission and objectives statements, it is suggested that the original statements demonstrate an early recognition of what are now acknowledged as key elements in a natural capital framework. For instance, acknowledging the value of water (and land) in the quality of life and economy of Calgary represents a fundamental first step toward protecting and sustaining natural capital. Tracking indicators of natural capital health through water quality data represents an important measure of human impacts on the environment, while reviewing and commenting on all project proposals that affect the river reflects a commitment to managing and investing in natural capital. Similarly, maintaining co-ordinated input to development, use and conservation of the city's watercourses and adjacent lands links directly with the need for co-ordinated policies and practices in a natural capital framework.

In 2000, on its tenth anniversary, the River Valleys Committee conducted a review of its mandate, performance and structure. The review summarised the many valuable and significant contributions made by the RVC, but pointed also to continuing concerns regarding effectiveness and efficiency that related directly to lack of funding. Although City Administration had established the RVC, and was to cover initial administrative and organisational set-up costs, it was the Parks Foundation, Calgary, which fully funded the RVC's operation.

In April, 2002, the City of Calgary's Standing Policy Committee on Operations and the Environment recommended that Council: (a) confirm the mission of the RVC; (b) direct the RVC and City Administration to continue to refine the RVC's development review role, processes and procedures; (c) direct the RVC to develop a detailed business plan and performance measures; and (d) direct the RVC's request for additional funding be brought forward through the 2003 budget process, for Council's consideration along with the detailed business plan and performance measures. Council accepted these recommendations and provided CDN$25,000 in short-term funding to support the business planning activity and the retention of the Parks Foundation, Calgary staff position related to the RVC. The RVC's Planning Steering Team prepared the business plan that subsequently was endorsed by the Executive Committee (River Valleys Committee 2003).

By May 2004, the RVC (with the Parks Foundation, Calgary and the City) had signed an operating agreement that would pay an operating grant to the Parks Foundation, Calgary for specified RVC services provided to the City. While payment for services represented a new way of operating for the RVC, the services to be provided related directly to the four roles that the RVC has undertaken since its inception. These re-named roles are:

a. community champion of river valley protection and recreational use – providing independent advice to City Council and Administration regarding a broad range of river and river valley policy and planning issues;
b. planning and project development review – assisting City business units by coordinating community (citizen and non-governmental organisation) input into a variety of policy, planning, development and project review processes;
c. public education and awareness – continuing to promote stewardship through training seminars, public presentations, river tours, and assisting the City with development of an annual education/awareness plan; and
d. participation in special projects and provision of special services – using the RVC's extensive membership and network to support specific projects (City of Calgary, Law Department 2004). Members of the Executive currently continue to work with the City advisors regarding full operationalisation of the agreement.

During its existence, the RVC has evolved into a vital citizen-driven organisation whose efforts to protect Calgary's natural capital have made a positive difference within the community and government. Although it is difficult to assess the value of the RVC's impact, it is clear that by 2003 the RVC had contributed well over CDN$7.5 million to the region in actual projects and related volunteer labour (River Valleys Committee 2003). While community good will and the value of the RVC's distinctive role as a community catalyst are more difficult to quantify, these contributions have occurred because of the RVC's unique position as a quasi-internal non-governmental organisation to the City of Calgary. This unique role as partner to government, community organisations, other non-governmental organisations and individual citizens has enabled the RVC to facilitate many projects that have added to the overall stewardship of Calgary's important waterways and valley parkland (River Valleys Committee 2003).

Acting in partnership with the City of Calgary, the RVC has helped progress toward the mutual goal of stewardship of the natural capital assets that Calgary's river valleys represent. Since the City has identified several corporate priorities (such as reducing Calgary's downstream impact on the Bow and Elbow Rivers) whose success will rely partly on the RVC's involvement, the RVC "remains well positioned to continue with this role in the future" (River Valleys Committee 2003, Attachment 2, 1). Calgary's municipal government administrators realise that operational and capital sustainability becomes more viable through community-based partnerships such as the RVC. For instance, in 2000 the RVC designed (with City support), produced and hand-delivered to more than 750 shoreline property owners, copies of the *Caring for Shorelands along Calgary's Waterways* brochure, an educational project that assisted the shoreline property owners to better manage and protect Calgary's waterways. As opportunity permitted, recipients were provided with a brief educational explanation of the intent and content of the brochure. The RVC's shorelands brochure project received CDN$10,000 from the City, and was valued at approximately CDN$27,000, providing the City with a positive return on its investment. The critical grassroots educational activities of the RVC also may be seen in the 2001 Bow River Shorekeepers Tour when seventy local shoreline property owners floated and paddled down the Bow River in rubber boats supplied by the RVC. The volunteer experts in riparian habitat management who accompanied the shoreline property owners assisted them in becoming even better stewards of the Bow River (benefiting Calgary and downstream communities).

Another dimension of the RVC's public education efforts are the forums the RVC hosts on topics ranging from stormwater management and riparian shorelands to City pesticide use and the *Water Resources Act*. These RVC forums make an important contribution to the community by bringing together a diverse and representative group of stakeholders, including local experts who may contribute as speakers, to network and discuss various issues and strategies outside of formal processes and in a non-confrontational manner. Not only does the RVC use the insights developed in these forums in its responses to City land use and development circulations (sometimes including challenges to the status quo and accepted direction of planning), but also the good will and positive community relations developed, as well as the empowered and responsible citizenry that result, help embed natural capital values directly into public policy decisions. These outcomes are invaluable contributions to community and government, and contribute to quality of life and economic growth in Calgary as well as to strengthening the natural capital/public policy framework.

The RVC's vision for the future of Calgary's river valleys indicates that the City and its residents will always be proud stewards of their rivers and river valleys, and will protect and nurture these key natural capital assets (principally from an ecological perspective), and demonstrate exemplary water quality management and conservation practices. The vision, featuring watershed and ecosystem-based plans and policies, anticipates continuous, integrated open spaces and parks on both sides of the rivers, responsible recreation, and sensitive urban development with appropriate set backs. In this anticipated future, citizens, their non-profit groups, corporations and governments will come to appreciate fully the diverse, interconnected values of the river valleys and will be "committed to collaborative planning and partnerships in pursuit of this common vision" (River Valleys Committee 2003, 3).

In looking to the future, the RVC recognises the many challenges it has yet to overcome. A key challenge is to meet the expectations that the City and others hold regarding the contributions of the RVC. There are a great many river and river valley related issues to which the RVC might provide input, yet maintaining credible outputs (such as timely responses to development applications) is difficult when the capacity to respond in a fully effective manner is constrained by lack of resources and working volunteers. RVC members recognise the wisdom and the potential given to them through the 1990 mission and terms of reference; members also know that the RVC's potential to act as informed advisor to City Council and Administration has only just begun. While the RVC model created by the City of Calgary enables volunteers to become involved and contribute at the strategic policy level, the process of involving volunteers at this level in city governance requires tolerance and encouragement by decision-makers and the development community. To their credit, Council and its various committees and some developers have supported the RVC's efforts to raise the level of debate on issues, and to contribute to a better balance among the environmental, heritage, aesthetic, social and economic values of Calgary's rivers and river valleys. The co-operative team approach that the City and RVC partnership represents allows for discussion of, and promotes change and improvement in, planning processes. Ultimately, this has raised the importance and understanding of natural capital values in the public policy arena.

Concluding Comments

As an important voice for the long term sustainability of Calgary's river valleys, including river reaches outside the city's boundaries, the RVC continues to occupy a unique role that brings benefit to community and government. The RVC's record of accomplishments reveals that, as an ambassador for education, it has delivered educational initiatives that directly support the City's priorities (and others) that the City likely would not provide because of cost. As a community voice, the RVC represents a wide variety of community-based organisations and benefits both the communities and the governments that need to access and work with each other. Through its educational events and forums, the RVC has developed the credibility to speak for the affected communities on matters pertaining to Calgary's river valleys. As a voice for the future, RVC members participate in the development of policy, legislation and standards, contributing directly to a public policy framework that protects and enhances such natural capital matters as recreational opportunities, land use, development, park and river management, and water quality that will contribute to a better future for residents of Calgary and its downstream neighbours. As a promoter of stewardship values and responsibilities, the RVC has helped (through its forums and educational events) to develop a strong sense of community responsibility for the river valleys and, over time, has seen a greater awareness of issues related to protection and enjoyment of the river valleys. Concurrently, the Committee's role within the land use and development process reinforces the RVC's voice for responsible protection and use of the natural capital that rivers, valleys and parklands represent.

As a catalyst for action, the quasi-non-governmental organisation River Valleys Committee is well positioned to bring together members of community and government to achieve the desired integration of stewardship of natural capital with economic growth. In this respect, the RVC may be considered an important model for other communities. For the past fifteen years, the RVC's presence and contributions in Calgary have helped improve the effectiveness of community-government relations and have used critical leverage to realise more of what is possible with the natural capital water assets in Calgary and southern Alberta.

References

Alberta Environment, Bow River Basin Council (2002). *Guidebook to Water Management: Background Information on Organizations, Policies, Legislation, Programs and Projects in the Bow River Basin. Prepared for the Bow Basin Plan: A Water Management Strategy for the Future of the Bow River Basin*. Edmonton: Alberta Environment.

Alberta Finance (2003). *Budget 2003*. Edmonton: Government of Alberta.

Berdahl, L. (2003). *Looking West 2003: A survey of western Canadians*. Calgary: Canada West Foundation.

City of Calgary (1984). *Calgary River Valleys Plan – The Plan and Policies,* Calgary: City and Community Planning Division.

City of Calgary (1990). *Supplemental Report CS 90-44-01 Attachment 2. Re: Public Participation in the Development and Maintenance of the Calgary River Valley System*. Calgary: Calgary City Clerks Office.

City of Calgary (2002). *Our Environmental Stewardship: Special report to citizens*. Calgary, AB: City of Calgary.

City of Calgary (2003a). *Striking a Balance: managing Calgary's continued growth*. Calgary, AB: City of Calgary.

City of Calgary (2003b). *Finding the Balance: sustainable growth, equitable financing*. Calgary, AB: City of Calgary.

City of Calgary, Waterworks (2004). *Our Demand is Growing*. www.calgary.ca/docgallery/BU/cww/water_consumption_graph.pdf (accessed July 2004).

City of Calgary, Law Department (2004). *Memorandum of Agreement (#8053): Operating agreement between the City of Calgary and the Parks Foundation*. Calgary, AB: City of Calgary.

Corbett, B. and Lalonde, K. (Eds.) (2004). *The State of Southern Alberta's Water Resources*. Lethbridge: Alberta Agriculture, Food and Rural Development.

Davis, C. (2000). *2000 Review: A report on the function and future of the River Valleys Committee at the time of its tenth anniversary*. Calgary: Parks Foundation, Calgary.

England, R.E. (1998). *Natural Capital and the Theory of Economic Growth*. Durham, New Hampshire: Richard W. England Center for Business and Economic Research, University of New Hampshire.

Golder Associates Ltd. (2003). *Report on Strategic Overview of Riparian and Aquatic Condition of the South Saskatchewan River Basin*. Calgary: Golder Associates Ltd.

Harras, T (Ed.) (2003). *Supply, Quality and Use of Water in the Prairie Provinces*. Regina, Saskatchewan: Keewatin Publications.

McGuinty, D.J. (2002). *Natural Capital – A critical basis for improving Canada's standard of living*. Ottawa: National Round Table on the Environment and the Economy.

Nikiforuk, A. (2004). The big melt, *Explore*, 129, 34–38, 66, 68, 70.

Parks Foundation, Calgary (2004). *Wetlands: at work for you*. Calgary, AB: Parks Foundation, Calgary.

River Valleys Committee (2003). *Business Plan for the River Valleys Committee of Parks Foundation, Calgary*. Calgary, AB: Parks Foundation, Calgary.

Worbets, B. (2005). Having our cake and eating it too: investing in natural capital, *Dialogues*, 1, 2–3.

Worbets, B. and Berdahl, L. (2003). *Western Canada's Natural Capital: Toward a new policy framework*. Calgary, AB: Canada West Foundation.

8. A PICNIC IN MARCH: MEDIA COVERAGE OF CLIMATE CHANGE AND PUBLIC OPINION IN THE UNITED KINGDOM

TAMMY SPEERS

Introduction

The media play an important role in representing the environment in highlighting issues of concern via news and documentaries and also for showing the beauty of our natural environment through travel pages and programmes. Numerous large public surveys have found that the public depend on the media as their main source of science information which includes environmental issues (see, for example, Nelkin 1987; Einsiedel 1990; INRA Europe 2000; Office of Science and Technology and The Wellcome Trust 2000; Ministry of Research Science and Technology 2002; Hargreaves *et al.* 2003).

This chapter examines how climate change is represented in the British media and consumed by the public. It derives its data from an Economic and Social Research Council Report (Hargreaves, Lewis and Speers 2003). Newspaper, television and radio coverage of climate change is analysed in addition to two national surveys of the British public. The audience analysis reveals that repeated messages are more likely to be remembered by an audience but that this can lead to both understanding and misunderstanding. In 2002, at the time the media and audience analyses were carried out, climate change was in and out of the news. During that year, springtime floods hit parts of the UK; between January and March the Antarctic ice-shelf 'Larsen B' collapsed; later in the year the UK government signed up to the Kyoto protocol and in August the second Earth Summit was held in Johannesburg, South Africa.

Since 1980 academics have carried out considerable analysis of the media coverage of environmental issues such as climate change. Media scholars have concluded that environmental issues often do not receive sustained coverage because they are not event oriented and instead are based on long-term events (Allan *et al.* 2000, 9). Coverage of environmental stories can, like many other long-term stories, be cyclical, as Einsiedel and Coughlan (1993) claim was the case with coverage of how climate change was increasing in the late 1980s and moved in and out of the news agenda.

G. Humphrys and M. Williams (eds.), Presenting and Representing Environments, 121-135.

The late 1980s are commonly regarded as the heyday of media environmental coverage with the employment of specialist environmental correspondents becoming commonplace in media outlets. By the early 1990s environmental stories were dropped from news agendas "when its importance became more generally accepted by industry and politicians" (Gee 2000, 212, see also Allan *et al.* 2000, 9). Indeed, an analysis of television news in the early 1990s found environmental stories made up less than 5 per cent of the news in BBC (British Broadcasting Corporation) and ITV (Independent Television) daily broadcasts (Cottle 1993). Coverage of climate change has similarly evolved as environmental stories moved down the news agenda.

Moorti describes the changing emphasis in climate change stories: "Global warming reporting began with catastrophism, with dramatic overstatements at the beginning of a new finding as a news hook, and now uncertainties are being overstated as part of a new hook to a story that is generally accepted" (in Wilson 2000, 205, see also Ungar 1995). In addition, the story about climate change is often a story about government policy and this can be difficult to make newsworthy. "The need to report dramatic news has proved an obstacle to discussion of the slow unravelling of the science and policy debates surrounding global environmental change issues" (Smith 2000, 1). Journalists and, more importantly, their editors, regard climate change as a story that has 'already been covered' thus it can no longer receive the sustained coverage it received in the late 1980s when it first came to attention outside the scientific community.

When Bell studied levels of accuracy about climate change science in newspaper articles he found that journalists were more likely to exaggerate rather than reduce measurements. For example, journalists were more likely to exaggerate and not underestimate the rise of sea-levels. He concludes that "inaccuracies are at base caused by the news values which journalists ... work by" (1994, 271). The emphasis on bad news, on the potential of horrific events happening, is more newsworthy than information discussing potential subtleties over a longer period.

Audiences reflect the intermittent coverage given to climate change. Agenda-setting theory, the idea that the media may not be successful in telling people what to think, but are more successful in telling people what to think *about* (McCombs and Shaw 1972), is often used to demonstrate the power of the media over audiences. So-called agenda-setting studies have suggested that media coverage does influence public attention to climate change issues (see, for example, Trumbo 1996; Hester and Gonzenbach 1997). More commonly though, public opinion polls/studies of climate change have measured the level of concern of the public and find they are concerned about climate change (see, for instance, Taylor 2000; Poortinga *et al.* 2003). Public opinion polls generally measure opinion rather than knowledge, but the Eurobarometer studies measure both knowledge and opinion. Their most recent study of the European public found the public confused the ozone layer with the effects of climate change but three-quarters of the public correctly stated that the greenhouse effect "can make the sea level rise" (INRA Europe 2000, 26). Wilson (2000) also surveyed levels of knowledge of US college students and environmental correspondents and found that the students were confused about climate change. The UK government measured public knowledge of climate change in 2000. They found the respondents confused about the causes of climate change and that from a list of causes they were most likely to name "change in temperature and weather patterns in the UK" and less likely to identify other effects such as "loss of habitats", "more frequent and stronger storms" or "risks of new diseases" (DETR 2000). The media coverage emphasises environmental changes, indeed the term 'climate change' itself emphasises this aspect of global warming, but the subsequent effects rarely receive substantial media attention. Pictures can "set the television agenda of a greenhouse world" (Wilson 2000, 206).

Method and Sample

This chapter examines the media coverage of climate change during a seven and a half-month period in 2002. The methodology is two-fold: sample content analyses of television, radio and newspapers together with two nationally representative surveys of public opinion and knowledge.

The media content sample analyses include television, radio and both the broadsheet and tabloid newspapers. The sample covers the period from 28 January to 15 September 2002 inclusive. The weekday BBC 6:00 evening news and ITV 6:30 evening news were selected for inclusion as they are the two news programmes with the largest audiences in the UK. The newspaper sample (using the Lexis-Nexis database) consists of five dailies: *The Guardian, Daily Telegraph, Daily Mail, The Sun, Daily Mirror,* and four Sunday newspapers, *Mail on Sunday, Sunday Times, News of the World* and *The Observer*. The BBC radio news programme *Today* (highlights archived on the Internet) was selected because of its large (albeit rather 'up-market') audience for a radio news programme and because of the influential role it plays on the wider news agenda. Highlights archived on the Internet (www.bbc.co.uk/radio4/today) were analysed. (When asked how it was decided which stories would be included, an email from the *Today* programme responded that they simply selected the 'significant' stories.)

All those items that contained the terms 'climate change' or 'global warming' were analysed in detail. The stories were then divided into four categories: primary theme, secondary theme, single-mention and letters. The criterion for inclusion in the primary theme group was discussion of climate change in more than one or two sentences. Of the 829 items located, 255 had climate change as the central theme and 159 as the secondary theme. Unless otherwise stated, these newspaper, television and radio items are the subject of the content analysis (n = 454) and are all merged together in the category of 'main and secondary theme stories' (see Table 8.1). The letters to the editor, regardless of the centrality, were categorised separately and were not included in the primary theme category. As Table 8.1 shows, most references to climate change tended to be a secondary or incidental part of the story.

The two surveys were carried out in April and October 2002. Both surveys were conducted in ten cities around the UK. A total of 1035 interviews were carried out in

April and 1037 interviews in October and spread evenly between the locations above. The surveys were conducted by Research and Marketing Ltd. in Cardiff, Wales, UK and they also carried out the statistical analysis according to the research team's requests. All interviews were conducted on the street around the busy areas of the city and town centres. Each survey is nationally representative of the national average in terms of gender, age and social class. For all data presented here $n = 1035$ in the April survey and $n = 1037$ in the October survey.

Table 8.1 *Environmental stories and the media*

	Main theme	Secondary theme	Single mention	Letter	Total
Broadsheets	149	106	286	38	579
Tabloids	84	43	71	8	206
Television	11	7	5	N/A	23
Radio	11	3	7	N/A	21
					829
Percentage of all stories (%)	31	19	44	6	

Media Coverage

As Table 8.2 shows, there was considerable difference in the amount of attention paid to climate change by the different media with broadsheets far more likely to cover the story. The total television news coverage on the UK's most watched bulletins, the BBC and ITV early evening news, over the seven and a half month period amounted to fifty-eight minutes and on the Today programme fifty-nine minutes. *The Guardian*, Britain's left-wing broadsheet, published more than twice as many climate change stories as any other paper in the sample, while tabloids like *The Sun* and its sister Sunday paper, the *News of the World,* gave the story very little coverage at all. This may reflect the greater emphasis the broadsheets have in covering international issues, as well as *The Guardian's* assumptions about their readers' interests and their practice in campaigning on left-leaning environmental issues.

Table 8.2 *Number of articles*

Daily Mail	79
The Mirror	78
The Sun	32
Daily Telegraph	133
Guardian	320
Observer	56
News of the World	3
Mail on Sunday	14
Sunday Times	70

Coverage of effects but not causes

Media studies scholars' attempts to systematise why journalists select some stories over others and why stories are reported in certain ways use the theory of 'news values'. Examining which stories are likely to be covered and the typical ways they are covered has led to concepts of what information or stories are 'valuable' or 'newsworthy'. Golding and Elliot (1979) state news values are used both to select material and act as guidelines for presentation. Few studies acknowledge that news values or concepts of newsworthiness might change over time in light of the evolution of the media. There are periods when certain stories are more 'valuable' than at other times, and frameworks for understanding stories can modify or change over time "but at any given moment they (frameworks) influence both the selection of news stories and the way in which the particular news ingredients of the story – the bits of information to be included – are assembled" (Manning 2001, 61). The media analysis found that certain news values influenced how the climate change story is reported. Stories about climate change more likely to be covered are those about climate change in the UK (proximity), stories that provide 'good' pictures or images (entertainment and novelty), stories that support a media outlet's own agenda – *The Guardian's* had consistent coverage of climate change (campaigns), and stories that affect a large number of people (magnitude) – for example, the floods in the UK.

While there are many references to actual or potential changes in the British climate, the specific effects on the rest of the world – particularly where the effects of global warming are likely to be devastating – receive much less attention. The main threat of climate change is seen in British terms – specifically, that an already fairly damp climate will become even wetter, notably in winter.

When climate change is the main or secondary theme or focus of the story, the *causes* or *effects* of climate change are mentioned more often than not. Overall, 58 per cent of stories in newspapers, television and radio refer to the causes of climate change and 65 per cent mention at least one effect (in 2002 flooding and a warm spring were most commonly discussed). Just over a third (34%) of stories mention both a cause and effect (Table 8.3).

Tabloid newspapers are less likely than broadsheets to mention causes of climate change, but more likely to refer to its effects (perhaps because the effects tend to be more dramatic than the causes).

Table 8.3 *Causes and effects of climate change and the media*

	Both cause and effect	Cause only	Effect only
Broadsheets	19	59	45
Tabloids	26	48	58
TV	50	61	83
Radio	43	64	71
Overall averages	34	58	65

Table 8.4 Answers (%) to the survey question: 'As far as you know, which of the following are causes of climate change?'

	April	October	Average
Destruction of forests	69.5	73	71
Nuclear Power Plants	44	44.5	44
Carbon Dioxide Emissions	64.8	67.4	66
Air Pollution	70.8	73.9	72
Hole in the ozone layer	66.4	64.4	65
Don't know	6.1	5.3	6

The causes of climate change are usually mentioned briefly and with little context (Table 8.4). So, for example, when causes of climate change are mentioned, the words 'fossil fuels' are regularly used but rarely defined. What is much less clear from the coverage is how fossil fuels or greenhouse gases might actually create global warming. The following example from the *Daily Mail*, written by science correspondent James Chapman, is a rare example of the media explaining *the link* between fossil fuels and climate change and a definition of the greenhouse effect:

> In a study to be published this month by the journal Geophysical Research Letters, a Belgian team used computer models to analyse the effect of adding one per cent more CO_2 to the atmosphere each year over 70 years. CO_2 levels have been rising for decades, thanks to traffic, fossil fuel consumption and industrialisation ... Many believe the rising levels of CO_2 in the atmosphere are trapping more of the sun's heat, causing global warming. (*Daily Mail* 13 February 2002)

In fact, Chapman's coverage in the *Mail* was alone responsible for nearly a third of all the explanations of the greenhouse effect provided in the sample of newspapers. It is also a good example of how tabloid newspaper coverage does not have to involve 'dumbing down': in this case a tabloid journalist uses his skill for conveying information simply and economically to both explain and inform. Explanations of greenhouse gases are rare – while *greenhouse gases* of one form or another (generally carbon dioxide) are mentioned in 39 per cent of newspaper articles, the *greenhouse effect* itself is only defined in 3 per cent of newspaper articles (Table 8.5). Thus proving that the details of climate change are not considered quite as newsworthy as its effects or causes.

Table 8.5 Answers (%) to the survey question: 'Which of the following explains the way that greenhouse gases effect our climate?'

	April	October
They directly warm the earth's surface	13	17
They thin the ozone layer	54	53
They prevent heat from escaping the earth's atmosphere	16	17
They evaporate cloud cover	2	2
Don't know	15	12

Single-reference

Close to half (44%) of all references to climate change in the media sample are incidental; single sentence references contained in stories about another topic. Climate change is often referred to in connection with other environmental and weather related stories. The inclusion of climate change in such a list tends to reduce it to simply another environmental problem and thus its importance can be too easily dismissed or forgotten. For example, in a BBC television news report on flooding in Bolivia, the cause of the flooding is stated in the following sentence whilst the effects make up the rest of this two-minute report:

> The government says this was an unpredictable act of nature but such scenes are all too familiar in Latin America with a combination of deforestation and climate change making many cities vulnerable.
>
> (*BBC Television News* 21 February 2002)

Climate change is also often included in stories about political differences between Europe and the United States (US) (specifically, Europe's support and US opposition to the Kyoto protocol). The following examples are fairly typical:

> The Bush administration has already pulled out of the Kyoto Protocol on global warming, and made it clear that it no longer considers itself bound by the Anti-Ballistic Missile Treaty. It also denies that hundreds of prisoners taken from Afghanistan to the US naval base in Cuba, Guantanamo Bay, are covered by the Geneva Convention.
>
> (*Daily Telegraph* 7 May 2002)

> In recent months, the American rogue state has torn up the Kyoto treaty, which would decrease global warming and the probability of environmental disaster. It has threatened to use nuclear weapons in "pre-emptive strikes" (a threat echoed by Hoon). It has tried to sabotage the setting up of an international criminal court, understandably, because its generals and leading politicians might be summoned as defendants.
>
> (*The Mirror* 4 July 2002)

Part of the reason for the high number of single-mentions of climate change is because news editors consider it as a story that has been covered. Specialist environmental correspondents have complained that it is difficult to convince their editors to run stories about climate change. One former environmental correspondent explains how he managed to get a story about climate change on the front page: "The story came to me as a global warming story but I managed not to mention this until well into the copy" (Brown and McDonald 2000, 67).

Another reason why there are so many single-references to climate change is that the issue has moved on from being only a 'hard news' story to being a 'soft news' story. Scientists often accuse journalists of only covering science stories that can be made to have a human angle (see, for example, Hansen 1994, 114). These types of stories, labelled 'soft news', are included in the news in order to entertain the audience. For example, the anticipated effect of the UK having a warmer climate as a result of climate change is seen as not so detrimental (though not always unproblematic). In a Sunday broadsheet, one food writer welcomes the possible benefits climate change will bring:

> Listen – I am as worried as anyone else about global warming, but having lunch outdoors five times by the middle of March has been a bit of a treat.
> (*The Observer* 14 April 2002)

The Sun, Britain's most popular newspaper, used a climate change story as an opportunity for some of its renowned *double entendre* when referring to the decline in tits, a common garden bird in Britain. The headline reads "WHERE ARE ALL THE TITS? – Figures droop as lots of Little Bustards nip in" and the article begins:

> TITS are on the decline in Britain, *The Sun* is sad to reveal. Far fewer of the popular birds are seen in public because of our changeable climate, scientists say. But foreign species like the Little Bustard are on the rise, lured by our sunny skies and blown in on strong winds.
> (*The Sun* 4 April 2002)

Indeed, this story was so irresistible that *The Sun* published an investigative follow up the following day under the headline: 'TITS OUT FOR OUR LAD'.

Whilst Britain lacks the extreme change of seasons typical in countries like Canada, they are nonetheless just as pleased when the warm air returns in the spring. Many stories welcomed the spring temperature which appeared in January and February in 2002. One gardening article in the *Daily Telegraph* concluded, "Global warming has its upside" (20 April 2002). Early in the year ITV news ran a jocular two-minute item about the increasing temperatures. The general news correspondent admitted that "warm weather doesn't always bring good news" but the most serious consequence of the warm weather was the ruined floral displays at Kew Gardens in London. Although the UK would later be hit with severe spring-time flooding, this piece failed to include any serious effects of the warmer weather. The story ends in typically light-hearted fashion with the only effect of the warmer weather would be that new plants can now be planted in the UK:

> In the long term scientists believe the balance of nature may shift as plants that thrive in the warmth force out those which prefer the cold.
> (*ITV News* 19 February 2002)

Climate change is regarded as soft-news by some journalists, especially those who are not specialist news correspondents, because there are few 'new' angles or ideas to report. This is a story that broke in the 1980s and really has not significantly changed since then. As one BBC environmental correspondent admitted "Climate change *was* a great story" (Kirby 2002). The first assertions about the possible impact of climate change and the subsequent media attention made the Western public aware, and somewhat concerned, about this looming crisis. Journalists might be interested in covering climate change but, without new angles or interesting images, coverage is likely to be substantial only when newsworthy images, such as floods, are available.

Audience

The media analysis demonstrated that environmental change was an issue that received more attention in the broadsheets than the tabloids. It is notable, then, that in the October survey 83 per cent of *Guardian/Independent* readers felt that climate change "was something we should be concerned about", compared with only 55 per cent of *Sun* readers. Previous research has shown that "Mass media portrayals of environmental information are known to be effective mostly at the upper-rungs of the socio-economic ladder, for instance with underprivileged classes receiving relatively little information" (Bailey in Comas and Shanahan 1999, 35). A significant proportion of readers would therefore appear to be following their newspaper's lead in their levels of concern or else have selected a newspaper in line with their own preoccupations.

Concern Reflects Media Coverage

More profound, perhaps, is the way media content is generally reflected in the public understanding of climate change. In short, people are aware of the issue, and most people are able to link climate change to the destruction of forests and carbon dioxide emissions. But few people can explain the process behind this link, and, as a consequence, see climate change as a consequence of a whole medley of environmentally sensitive issues.

As the media analysis demonstrated, media coverage does pay attention to the proposed causes of climate change with 55 per cent of stories making some reference to the causes, but only in a cursory way. It is perhaps not surprising then, that confronted with a list of possible causes, most people tend to respond with a rather eclectic list. Respondents could choose from a list of matters that are of environmental concern, but some have little to do with climate change.

On the surface, the number of people who see the hole in the ozone layer or nuclear power plants as causes of climate change is difficult to explain: this is not, after all, a claim made in media coverage. But the fact that so many people make these links tells us something about the way people learn from the media. In the case of climate change, this involves turning associations (between environmental problems like the thinning ozone layer and global warming) into causal links. In the last few years, media reports have often mentioned global warming and the ozone layer in the same breath:

> WHEN British engineers pioneered the industrial revolution and the world's first great factories 200 years ago, their eye was on progress and

> profit – not pollution. But scientists now believe it put us on a collision course with nature, which will see the destruction of many aspects of life on earth. Hundreds of millions of cars now spew out poisonous gases. Cities belch smoke up into the ozone layer, while distant conferences sit and discuss what to do about it. At the heart of the debate is the effect of "global warming" and the "greenhouse effect", caused by us producing too much carbon dioxide from burning fossil fuels. If addressed quickly enough, we might start to slow the process of the planet getting hotter.
>
> (*Daily Mirror* 29 March 2002)

This example does not assert a causal connection and yet, in the absence of any other explanation offered, most people tend to assume one. This is suggested more clearly by the response to a question asked in both surveys about the greenhouse effect. The erroneous idea that greenhouse gases affect the climate by thinning the ozone layer outnumbers correct responses (they "prevent heat from escaping out of the earth's atmosphere") by more than three to one.

Bell also found that people often attributed rising temperatures to ozone depletion, which he describes as a "classic case of meltdown between events with similar mental scripts" (1991, 245). The most recent Eurobarometer report also found that 70 per cent of the European public surveyed claimed to understand climate change but nearly 56 per cent erroneously believed that holes in the ozone layer will cause more storms and tornadoes. The greenhouse effect is rarely explained in media coverage.

It could be argued that ignorance of the technicalities is less important than a more holistic understanding of the environment. For example, it could be argued that people are right to understand the thinning ozone layer *and* the greenhouse effect as linked, in the sense they are both environmental problems caused by industrialisation and human activity. Since both have negative effects, confusion of what causes which might be less important than a more general awareness of the environmental consequences of our actions. This issue will be addressed in the conclusion, but here what is interesting about these responses is what they reveal about the way media coverage (whether it means to or not) creates explanatory frameworks.

Interestingly, the age group *most* likely to get this particular question right was the one with the lowest scores overall – the 18–25-year-olds. Those in this age group with more education did particularly well, 22 per cent of those with A-Levels and 37 per cent of those with a degree answering correctly. This suggests that schools have recently included a scientific explanation of global warming in their teaching, with a considerable degree of success.

The media analysis suggested that the news media gives considerable attention to the consequences of global warming when it affects the UK. But instead of asking about 'global warming', in order to test how widespread knowledge of other consequences was, the survey asked a question with a correct response that is in some ways counter intuitive. The predicted increase in winter rainfall might seem, for some, to contradict the idea of 'global warming' (for British people, wet weather and cold

Table 8.6 *Answers (%) to the survey question: 'Which of these is a predicted climate change for the UK?'*

	April	October
Less rainfall in winter	19	24
More rainfall in winter	53	52
No effect on rainfall	5	4
Don't know	23	19

weather are often seen to go together). For the amount of coverage given to the springtime flooding in 2002, it is perhaps surprising only half of the public could correctly identify that rainfall will increase (see Table 8.6).

If the news tends to pay less attention to the processes behind global warming, it appears from Table 8.6 to be more instructive in creating an association in people's minds between climate change and increased rainfall. This fits, once again, with news values, in which the dramatic flooding in the UK in recent years has received substantial coverage.

General lack of certainty about the causes of global warming is also reflected in the difficulty people have in connecting the local with the global, thereby understanding how the daily choices in their own lives might be linked to climate change. So, for example, although a number of stories linked global warming to the phrase 'fossil fuels', few explained what this meant, and less than 2 per cent of the articles mentioning climate change in the study made any reference to how the public might contribute to reducing the rate of climate change. A rare attempt in the sample to deal with this type of issue is made by a travel editor in a Sunday broadsheet:

> [O]ne seat on a flight from the UK to the US is responsible for as much global warming from carbon dioxide emissions as one average driver using a car in the UK for a year. (*The Observer* 17 April 2002)

While it is very difficult to be definitive about environmental impacts, airplane emissions which burn fossil fuels when flying goods around the world contribute more to climate change than buying produce transported locally. Since this is a good example of how people's everyday choices may have an impact on the issue, as well as one which reveals a degree of understanding of what fossil fuels are, we asked the public to make such a choice (See Table 8.7).

Table 8.7 *Answers (%) to the survey question: 'In a supermarket, which of these choices would have the* LEAST *impact on climate change?'*

	April	October
Buying organic apples from New Zealand	26	23
Buying non-organic apples that are locally produced	40	44
Don't know	34	34

The word 'organic' was deliberately used to create confusion as it is generally associated with environmentalism. The level of incorrect answers suggests once again that for around a quarter of the sample, associations may be standing in for causal relationships. The answers do suggest, nonetheless, that more people (more than four in ten) were aware enough of the role of fossil fuels to make the hypothetical choice for local produce.

Climate change has become a major international political issue in recent years, and the surveys included a question to examine whether one of the main points of conflict in international negotiations on climate change is common knowledge (see Table 8.8).

Table 8.8 *Answers (%) to the survey question: 'The Kyoto Protocol aims to get countries to agree measures to reduce the threat of climate change. Which of the following is opposed to the Kyoto Protocol?'*

	April	October
European Union	3	4
USA	52	48
UK	2	1
Mexico	4	3
Don't know	40	45

In 2002 the Kyoto protocol received a moderate amount of media coverage and around half the respondents correctly answered the question. What is more notable about this response is the comparative absence of incorrect assumptions. This appears to be one of those areas that people either feel they know or they don't – over 90 per cent of responses are either correct or 'don't know'. The fact that this is unusual is indicative of the process by which people learn from news: this is one of the few questions that did not involve answers where associations people draw from media coverage might steer them in the wrong direction.

Overall, the relationship between media coverage and public understanding regarding climate change demonstrates both the media's power to inform and, inadvertently, to misinform. Most people have clearly learnt a number of things about climate change from the media but the way people absorb information from the news means that they are also inclined to mistake thematic links for causal links. This, in turn, has made people concerned about climate change, but they are much less clear how it works. Research on the relationship between news and public understanding suggests that most people consume news rather inadvertently, and hence only take in certain aspects of a story. They also find it hard to follow the often truncated narratives of news – which tend to start with the climax of a story before giving any history or background – and hence a great deal of information passes people by. The information that *does* stick, in this context, is often based on oft-repeated associations (Lewis 2001). People then use what information they have to make sense of the world

– what has been called "low information rationality" (Sniderman, Brody and Tetlock, 1991) or "information short-cuts" (Popkin, 1991).

Discussion

This research demonstrates that certain repeated themes about the causes and consequences of climate change are filtering through to the public, although there is also a degree of confusion that appears to come from reporting two or more different environmental problems together. This may make it difficult for many people to judge the specific merits of climate change policy proposals. Some policy makers might argue, for example, that the connection many people make between nuclear power and climate change is fallacious and thus profoundly unhelpful. But one could also argue that if people are somewhat undiscriminating in their assessment of the relationship between environmental problems – from air pollution to nuclear waste – the rather muddled picture that emerges does have a certain coherence. Most people are able to make links between a range of human activities that have a negative environmental impact.

Media coverage of climate change often refers to its causes, and there is a fairly high degree of public awareness that human activities such as deforestation and fossil fuel emissions are said to cause climate change. What is less apparent in news coverage is a focus on the scientific process involved, such as the nature of the greenhouse effect. The absence of repeated explanations is reflected in public understanding: most people clearly do not know what the greenhouse effect is or how it works. Thus what matters is weight and pattern of coverage, not detail. People use information to construct an explanation and thus ideas with little media presence – such as the mechanics of the greenhouse effect – are unlikely to filter through and yet many people still feel able to make knowledge claims. They appear to do so on the basis of the presence of oft-repeated associations in media coverage: in this case, between the thinning ozone layer and the greenhouse effect. What is, in media coverage, merely a juxtaposition (under the general heading of human-made environmental problems) undergoes a cognitive leap in public understanding, so that it is understood as a causal relationship.

Themes or frameworks are then used as building blocks for people to make sense of an issue. This can, in practice, be a fairly crude cognitive process, and while these building blocks can be put together in ways that facilitate public understanding, many people ignore the fine print and assume connections between things simply because they are often juxtaposed in media coverage. Most people need practical rather than technical information. The widespread use of scientific shorthand – notably terms such as 'greenhouse gases' or 'fossil fuels' – *assumes* rather than *communicates* an understanding of the causes of climate change.

If there is a gap in media coverage and public understanding of climate change, it is not a simple absence of scientific understanding. The problem is more a question of emphasis. First, while the predictions coming from the United Nations committee on climate change (amongst others) are dramatic and catastrophic, most of the misery is likely to fall on countries in the third world. The effects of climate change on Britain are seen as far less alarming. While the media often discuss the consequences of climate change, media coverage on this issue – as on many others – is often somewhat parochial, with many casual references to its more benign effects. This may explain why the issue is not generally seen as important. This may also partly explain why broadsheet readers, whose newspapers have a more international focus, are more concerned about this issue. The challenge is to make the sheer scale of potential global damage that climate change may inflict, a major and recurrent news story. What matters is not so much the science itself, but establishing a clear connection between science, policy and the broader public interest.

References

Allan, S. Adam, B. and Carter, C. (2000). Introduction. In S. Allan, B. Adam and C. Carter (Eds.) *Environmental Risks and the Media.* London: Routledge.

Bell, A. (1994). Media (mis)communication on the science of climate change, *Public Understanding of Science,* 3, 259–275.

Brown, P. and McDonald, F. (2000). Have we 'had enough of all that eco-bollox'? In J. Smith (Ed.) *The Daily Globe.* London: Earthscan Publications.

Cottle, S. (1993). Mediating the environment: modalities of TV news. In A. Hansen (Ed.) *The Mass Media and Environmental Issues.* Leicester: Leicester University Press.

DETR (Department of the Environment, Transport and the Regions) (2001). *Digest of Environmental Statistics*. www.environment.detr.gov.uk/des/index.htm (accessed December 2001).

Einsiedel, E. (1990). *Scientific Literacy: A survey of adult Canadians.* Calgary: University of Calgary.

Einsiedel, E. and Coughlan, E. (1993). The Canadian press and the environment: reconstructing a social reality. In A. Hansen (Ed.) *The Mass Media and Environmental Issues.* Leicester: Leicester University Press.

Gee, D. (2000). Communicating complexity and uncertainty: a challenge for the media. In J. Smith (Ed.) *The Daily Globe.* London: Earthscan Publications.

Golding, P. and Elliot, P. (1979). *Making the News.* London: Longman.

Hansen, A. (1994). Journalistic practices and science reporting in the British press, *Public Understanding of Science,* 3, 111–134.

Hargreaves, I., Lewis, J. and Speers, T. (2003). *Towards a Better Map: Science, the public and the media*, an Economic and Social Research Council Report, available at www.esrc.ac.uk/publications (accessed November 2003).

Hester, J. and Gonzenbach, W. (1997). The environment: TV news, real world cues, and public opinion over time, *Mass Communication Review,* 22, 5–20.

INRA Europe (2000). *Eurobarometer 52.1: The Europeans and Biotechnology.* www.europa.eu.int/comm/research/pdf/eurobarometer-en.pdf (accessed November 2003).

Kirby, A. (2002). Climate change: what a great story it was. Speech given at Oxford University's Green College Centre for Environmental Policy and Understanding Conference. www.climatechange.org (accessed January 2003).

Lewis, J. (2001). *Constructing Public Opinion: How political elites do what they like and why we seem to go along with it.* New York: Columbia University Press.

Manning, P. (2001). *News and News Sources: A critical introduction.* London: Sage.

Ministry of Research Science and Technology (2002). *Commonsense, Trust and Science* www.morst.govt.nz/publications/commonsense.html (accessed November 2003).

McComas, K. and Shanahan, J. (1999). Telling stories about global climate change: Measuring the impact of narratives on issue cycles, *Communciation Research,* 26, 30-57.

McCombs, M. E. and Shaw, D. L. (1972). The agenda-setting function of mass media, *Public Opinion Quarterly,* 36, 176-185.

Moorti, S. (1991). *Newspaper Coverage of Global Climate Change by Five Papers*. College Park, MD: Center for Global Change, University of Maryland.

Nelkin, D. (1987). *Selling Science: How the press covers science and technology.* New York: W.H. Freeman.

Office of Science and Technology and The Wellcome Trust (2000). *Science and the Public: A Review of science communication and public attitudes to science in Britain.* London: OST.

Popkin, S. (1991). *The Reasoning Voter: Communication and persuasion in presidential campaigns.* Chicago: Chicago University Press.

Poortinga, W., Pidgeon, N. with contributions from K. Bickerstaff, I. Langford, J. Niewöhner, T. O'Riordan and P. Simmons (2003). *Public Perceptions of Risk, Science and Governance.* Centre for Environmental Risk, University of East Anglia. www.uea.ac.uk/env/pur/Final Report Risk Survey 2002.pdf (accessed November 2004).

Smith, J. (2000). Introduction. In J. Smith (Ed.) *The Daily Globe.* London: Earthscan Publications.

Sniderman, P, Brody, R. and Tetlock, P. (1991). *Reasoning and Choice: Explorations in political psychology.* Cambridge: Cambridge University Press.

Taylor, H. (2000). Most people believe global warming is a serious problem, but do not think it causes most forest fires. www.harrisinteractive.com/harris_poll/index.asp?PID=108 (accessed March 2002).

Trumbo, C. (1996). Constructing climate change: Claims and frames in US news coverage of an environmental issue, *Public Understanding of Science,* 5, 269–283.

Ungar, S. (1995). Social scares and global warming: Beyond the Rio convention, *Society and Natural Resources*, 8, 483–456.

Wilson, K.M. (2000). Communicating climate change through the media: Predictions, politics and perceptions of risk. In S. Allan, B. Adam and C. Carter (Eds.) *Environmental Risks and The Media.* London: Routledge.

9. CHALLENGING THE NEGATIVE CRITIQUE OF LANDSCAPE

ROBERT A. NEWELL

The critic William Packer indicated the adverse critical context of landscape painting in his catalogue introduction to the *Contemporary British Landscape* exhibition in 1999 at the Flowers East Gallery London. He remarked on "the simple truth that landscape painting is off the critical and curatorial agenda and has been for some considerable time" (Packer 1999, 3–4). This is a longstanding critical tendency and five years later Packer's observation still stands. What I call the Negative Critique of Landscape is a multi-disciplinary phenomenon – its impact on art criticism is but one associated aspect. The meaning of landscape is negotiated at the complex intersection of disciplinary concerns and issues, including geography, sociology, anthropology, artistic practice and aesthetics. The Negative Critique can be seen to be emergent from two major intellectual contexts. In one of these, articulated within Marxist traditions, landscape is regarded as an ideological construct and, in the other, environmental aesthetics, landscape is considered to be a pictorial conception of the environment that as such is considered to be detrimental to understanding and appreciating the multi-sensually experienced environment. As an artist, awareness of this critique sets up a tension with my own practice of landscape painting because this practice fundamentally concerns environmental presentation and representation. I have therefore felt compelled to investigate and address issues that this raises.

The Flawed Tradition – Ideology and Conventions

In recent years I have been making detailed drawings of rock formations. Sometimes, when I am out working on these, people come and have a look. They are probably tempted to think, or hope, that the drawing resembles the rocks it depicts. This is to manifest what is regarded in critical circles as the naive understanding of pictorial representation. In *Languages of Art,* Nelson Goodman (1976) radically challenges the role of resemblance in pictorial representation:

G. Humphrys and M. Williams (eds.), Presenting and Representing Environments, 137-151.

> The plain fact is that a picture, to represent an object, must be a symbol for it, stand for it, refer to it; and that no degree of resemblance is sufficient to establish the requisite relationship of reference. Nor is resemblance necessary for reference; almost anything may stand for almost anything else. (Goodman 1976, 5)

The Western practice of pictorial perspective is a particular target for Goodman's scepticism regarding resemblance. According to Goodman, perspective is a highly conventional system of representation that is considerably at odds with contemporary optics. For Goodman, pictorial representation is achieved through wholly artificial rules and conventions analogous to verbal language. According to this view, there is no natural or optical relationship of resemblance between picture and referent or object. It is important to emphasise that conventionality of this order operates within specific cultural groups. For the Negative Critique, the conventional language of landscape painting has an ideological basis that serves the interests of particular social elites in particular historical periods and geographical locations. Outside of these, it is redundant.

The word *landscape* itself can be used ambiguously. It can mean the environment *out there* or it can refer to landscape painting. The title of Kenneth Clark's (1979) *Landscape into Art* uses *landscape* to refer to the objective source for landscape painting. In *Landscape and Power,* W. J. Thomas Mitchell argues that the title of Clark's book, "*Landscape Into Art,* [is] quite redundant: landscape is already artifice in the moment of its beholding, long before it becomes the subject of pictorial representation" (Mitchell 2002, 14). He holds that the traditional conception of landscape is now redundant and exhausted; that its connection with painting makes it a source of falsehood, mystification and ideological control; and that our perceptions are consequently dominated by cultural conventions that sustain bourgeois and imperialist values.

Malcolm Andrews (1999) commences his *Landscape and Western Art* by using a photographic work of Karen Knorr's, *Pleasures of the Imagination: Connoisseurs* (1986), to illustrate what he argues to be the distanced, institutional, elitist and redundant status of landscape painting. Knorr's photographic compositions are both a celebration of the richness of museum and gallery interiors and a visual commentary on institutions of high-culture. In this photograph, landscape paintings become the signifiers of privileged bourgeois culture. These dark, flat, static, gilt-framed objects have little in common with the multi-sensual, dynamic environments of nature. They can be considered as the outsider's construction of landscape. Doubling the outsider's perception, the photograph invites us to observe the connoisseur's contemplation of the paintings with an alienated understanding, similar to that of overhearing a language one does not speak. Andrews concludes his introductory chapter by asserting that: "Landscape as a way of seeing from a distance is incompatible with [a] heightened sense of our relationship to Nature as living (or dying) *environment.* As a phase in the cultural life of the West, landscape may already be over" (Andrews 1999, 22).

The Negative Critique of Landscape may be summarised by the following propositions:

- "Landscape is an exhausted medium, no longer viable as a mode of artistic expression" (Mitchell 2002, 5).
- "Landscape … is an ideological concept" (Cosgrove 1998, 15).
- "Landscape is a particular historical formation associated with European Imperialism" (Mitchell 2002, 5).
- Landscape is a gender specific – masculinist formation (Rose 1993, 86–112).
- Current manifestations of the landscape tradition are likely to be debased kitsch versions of the redundant genre/medium (Mitchell 2002, 20).
- Landscape privileges sight, it constitutes a distanced *outsiders'* conception of the environment (Cosgrove 1998, 18, 24, 33). This visual basis of landscape has developed along with the conventions of perspective; perspective "glorifies the spectator" (Crandell 1993, 8).
- *Landscape* has dominated our conception of environment, it reduces this to a static, two-dimensional scene, inappropriate to its multi-dimensional, dynamic and evolving nature: "our experience of landscape is overwhelmed by pictures" (Crandell 1993, 166).
- Landscape painting and any aesthetics of environment derived from it lacks "evolutionary literacy", is scientifically uninformed: "It is superficial and narcissistic. In a word, it is trivial" (Callicot 1998, 134).

The occlusion of the natural by the conventional is visually encapsulated for Andrews by Rene Magritte's painting, *The Human Condition* (1933). This painting shows a landscape seen through a window, partially occluded by a painting that represents it. Andrews observes that. "Art here prevents us from seeing nature … or is *nature* simply an idea we need to construct for that which is *beyond* the painted, framed object?" (Andrews 1999, 127). Gina Crandell (1993, 162–165) illustrates the same point with Magritte's *The Door To Freedom* (1937) in her *Nature Pictorialised.* For many of the central exponents of the Negative Critique, landscape is precisely "the pictorial conception of nature" (Crandell 1993, 169).

Jean Baudrillard (1988) outlines a succession of stages through which the image in Western culture, has come to exclude reference to anything beyond itself:

> These would be the successive stages of the image:
> 1. It is the reflection of a basic reality.
> 2. It masks and perverts a basic reality.
> 3. It masks the absence of a basic reality.
> 4. It bears no relation to any reality whatever: it is its own pure simulacrum. (Baudrillard 1988, 170)

Mitchell argues that landscape can be understood in terms of commodity fetishism, the ideologically motivated occlusion of social reality by fetishistic substitution.

The commodification of landscape, and the development of its associated ideological functions in naturalising the values of the social elite, can be seen in the contexts of the property market, land management, the tourist industry, conservation, literature, the visual forms of painting, photography and film. Representation, replication, simulation; these have fuelled a tide of images and meanings that arguably can make no reference beyond their ideologically based meanings.

John Taylor (1994) in *A Dream* of *England* reproduces the photographic composition *Constable Country* by Paul Reas. This photograph is constructed around the theme of the directedness and schematisation of the tourist's experience and evaluation of landscape. A tour guide, dressed in a dark suit and looking rather like an estate agent, holds a reproduction of a Constable painting and points, directing the gazes of a tourist couple, by implication, at the view Constable painted: "Reas's work stands in the tradition of cultural commentators and historians who oppose the 'heritage industry' because they believe it promotes nostalgia and the ideology of conservation" (Taylor 1994, 247).

Claude glasses have become one of the material emblems of the eighteenth and nineteenth century Picturesque, a constellation of attitudes and values concerning landscape preferences. The Picturesque is a central target for the Negative Critique, it seems to dominate the environmental philosopher J. Baird Callicot's conception of landscape painting and associated aesthetics altogether. For Callicot, use of the Claude glass typifies the contrived indirectness of the Picturesque attitude to nature:

> A device of the period, the Claude glass ... tells the whole story. The new natural aesthetes carried the rectangular, slightly concave, tinted mirror with them into the countryside. Upon finding a suitable picturesque prospect, they turned their backs to it and rear viewed its image.
> (Callicot 1998, 133)

In *The Land Aesthetic,* Callicot (1998) roundly sets out a version of environmental aesthetics that is particularly hostile to the pictorial landscape tradition. This tradition, according to Callicot, cannot provide any basis for appropriate environmental appreciation:

> western appreciation of natural beauty is recent and derivative from art. The prevailing natural aesthetic, therefore, is not autonomous: it does not flow naturally from nature itself; it is not directly oriented to nature on nature's own terms; nor is it well informed by the ecological and evolutionary revolutions in natural history. It is superficial and narcissistic. In a word, it is trivial. (Callicot 1998, 134)

Callicot proposes that the *land aesthetic,* constitutes an "autonomous natural aesthetic" (Callicot 1998, 134) that "is sophisticated and cognitive, not naive and hedonic" (Callicot 1998, 139). He argues that the *land aesthetic* derives its sophistication from scientific knowledge, from "evolutionary literacy" (Callicot 1998, 135) that affords an informed way of experiencing an environment and understanding its variously authentic or violated characteristics. He further indicates the kind of judgement that this *evolutionary literacy* involves: "From the point of view of the land aesthetic, the attractive purple flower of centauria or the vivid orange of hawkweed might actually spoil rather than enhance a field of (otherwise) native grasses" (Callicot 1998, 135–136).

Attractiveness and *vividness* are aesthetic qualities irrespective of their evolutionary origins. A high valuation placed upon native species may or may not be based on aesthetic considerations. There is a problem here, of which Callicot seems unaware. Certain aesthetic qualities such as attractiveness or vividness have an irrefutable immediacy, their attribution in any instance is not deferred pending an investigation of their evolutionary native status. By contrast, what may be regarded as native species can turn out not to be so in the light of further knowledge.

The idea that a location's flora and fauna are native species could well be a factor in its aesthetic affectivity. Subsequent knowledge falsifying the attribution of native status may then create an aspect change, what we understood to be purely and authentically native might turn out to be there as a result of human intervention. In *Landscape and Memory,* Simon Schama (1996) cites an example of just such an aspect change to Yosemite National Park: "The brilliant meadow-floor which suggested to its first eulogists a pristine Eden was in fact the result of regular fire-clearances by its Ahwahneechee Indian occupants" (Schama 1996, 9). This aspect change does not, however, negate the aesthetic *brilliance.* Callicot fails to understand that valuing authenticity is not necessarily of specifically aesthetic relevance.

The environmental philosopher Allen Carlson similarly argues for an autonomous aesthetics of environment. Carlson contrasts this with what he terms the *Landscape Scenery Model* (LSM) of environmental appreciation:

> [LSM] construes the environment as if it were a static, essentially "two dimensional" representation, it reduces it to a scene or view. But the natural environment is not a scene, nor a representation, not static, and not two dimensional. In short, the model requires appreciation of the environment not as what it is and with the qualities it has, but as something it is not and with qualities it does not have. (Carlson 1998, 125)

The idea that the literally static two-dimensionality of painting restricts its meaning and content to static flatness is over-simplistic. Rudolph Arnheim (1966), particularly, has investigated the psychology of the dynamic field of forces constituted by the picture plane. The resources of pictorial language in general, and of painting and drawing in particular, are sufficiently rich to ensure that landscape painting is not the exhausted practice that its detractors maintain.

Claude Lorraine was one of a number of seventeenth century European painters who were highly esteemed in Britain, where their influence stimulated the development of the aesthetics of the Picturesque. Similarly to Crandell, Denis Cosgrove (1998) reduces Claude Lorraine's achievement to the establishment of conventions that helped to sustain landscape as ideology. In *Social Formation and Symbolic Landscape,* Cosgrove presents the thesis that the Western landscape tradition, following the Renaissance until its alleged virtual demise by the end of the nineteenth century, was dominated by Claude Lorraine's postulated *landscape idea.* This is defined as an ideological construct, the product of a specific historical process during which capitalism superseded feudalism. This was a period in which both of these socio-economic orders with their concomitant cultural superstructures were in states of relational tension, a dynamic disequilibrium whose inner contradictions were compensated for in a variety of symbolic representations or substitutions, most pertinently here, of course, by the *landscape idea.* Cosgrove's *landscape idea* has a "purely visual basis" (Cosgrove 1998, 32); in artistic terms it "stresses a personal, private, and essentially visual experience" (Cosgrove 1998, 14). It is the idea of the property-owning bourgeois spectator, projected on to the controlled and pacified environment that is reduced to the geometrical order of perspective centred upon that same spectator. It involves "the ideology of sight, distance and separation" (Cosgrove 1998, 270).

Insider/Outsider

The duality of *Insider* and *Outsider* relations with the land, formulated with considerable effectiveness by Cosgrove, extends John Berger's critiques of the painters Jean-Francois Millet and Thomas Gainsborough. Berger argues that Millet's paintings representing peasant labour are failures. His explanation for this failure resides in an alleged ideological dichotomy, that of the land inhabiting peasant (insider) and the artist bound by bourgeois cultural traditions and values (outsider). Thus, of Millet, he writes: "I believe that he failed because the language of traditional oil painting could not accommodate the subject he brought with him. One can explain this ideologically" (Berger 1980, 76). Berger's pronouncements upon the ideological roles of oil painting and of seeing possess an authoritative status for subsequent commentators, thus Cosgrove stated that: (my emphasis) "John Berger (1972, 83–108) has *shown* that oil painting is the bourgeois medium of painting *par excellence*" (Cosgrove 1998, 62). Berger declares that: "The claim of universal democracy was inadmissible for oil painting" (Berger 1980, 76–77). His ideological explanation really fails to materialise in his flawed analysis, for example, he offers this argument:

> [The oil paintings] fail because no unity is established between figures and surroundings. The monumentality of the figures refuses the painting. And vice versa. As a result the cut-out figures look rigid and theatrical. The moment lasts too long. By contrast, the same figures in equivalent drawings or etchings are alive and belong to the moment of the drawing which includes all their surroundings. For example, the etching of Going to Work, made ten years after the painting is a very great work ...
>
> (Berger 1980, 76)

Clearly the landscape backgrounds in the two painted versions of Millet's *Going to Work* (1851–1853) are subordinate to the peasant couple depicted walking through a field. At the time when these paintings were executed, Millet had never been particularly interested in either drawing or painting landscape directly. The human figure had been his main interest, particularly when labouring in urban or rural situations. After moving to Barbizon in 1849, however, the landscape context became, over time, steadily more important to him. Berger concedes that there are "differing degrees" of failure in Millet's paintings "which include figures" (Berger 1980, 76). Is there an ideological explanation for this variability? None is offered.

The etched version of *Going to Work* made in 1863, some twelve years after the oil paintings, shows the figures walking in an altogether different, and more realised landscape. This etching relates perfectly well to developments within Millet's oil painting. I suggest that it is the development over time that explains the difference between the etching and the oil paintings. Berger's ideological explanation merely introduces incoherent judgements that fail to explain anything at all about Millet or oil painting and etching. Arguably, Millet was notable precisely for the "unity [he] established between figures and surroundings" (Berger 1980, 76) and this can be conjectured as the ironic reason for the use of Millet by Berger to make his argument for the ideological culpability of oil painting.

If one looks at Millet's development of the theme of rural labour in his oil paintings, for example in *Man with Hoe* (c. 1863) painted at about the same time as the etching *The Walk to Work* (c. 1863), the relationship of people to the landscape within which they work is developed in very significant ways. The materiality of earth is combined with its extensiveness to create a challenging, resistant environment.

It is pertinent to compare Jules Breton's painting *The Gleaners* (1854) with Millet's version of the same subject (1857). Breton's painting is a scenic and distanced view of peasant labour in the fields – the receding plain of figures, the textural gradient of stubble fading towards the hazily sun-lit village punctuated by its church tower, all contribute to the pre-occupation with atmospheric perspective. This produces a pleasing picturesque interpretation of the subject, more plausible to argue as a scenic naturalisation of socio-economic inequalities.

The three women in the foreground of Millet's painting *The Gleaners* are bending over, searching the stubble and showing the signs of fatigue produced by the repetitious movements and postures that their labour entails. The nearest woman's head just coincides with the horizon without rising above it, producing an oppressive quality. Variations on the repetitive movements of the women are characteristic of the way Millet often dealt with the theme of human labour. The bulky haystacks and cart behind seem to weigh down on the two more stooped women in a way that negates what is often regarded as the inexorably distancing and neutralising effects of perspective. Millet and Breton both work with a conception of landscape and figures that is thoroughly perspectival. By Millet this is developed to establish an empathetic proximity with the gleaners, while by Breton it is used to distance the gleaners, more as optical incidents within a scenic rendering of atmospheric perspective. These differences are of fundamental importance in perceiving and understanding the two representations, and yet Berger seems to level such differences by imposing inappropriate ideological criteria, both paintings being identical in their *bourgeois* medium of oils.

Nonetheless, it is ideological analysis and derived judgements that continue to exert such a strong influence within the academic community. From a more feminist angle, Linda Nochlin considers that Breton and Millet are very much the same in their "assimilation of the peasant woman to the realm of nature" (Nochlin 1996, 32).

In a letter written in June 1885, Vincent Van Gogh drew a strong relationship between Millet's peasant labourers and the earth that applies just as strongly to the men: "How typical that saying is about the figures of Millet: 'Son paysan Semble peint avec la terre qu'il ensemence!' ['His peasant appears to be painted with the earth that he is sowing'] How exact and how true" (Roskill 1977, 227). Commenting upon "the strategies of his composition", Nochlin writes: "the bodies of the bending women are quite literally encompassed and limited by the boundaries of the earth itself: it is as though the earth imprisons them, rather than feudalism or capitalism" (Nochlin 1996, 32). This exclusively ideological reading of Millet's painting misses the larger and more appropriate conception held by Van Gogh.

One wonders if the *insider/outsider* distinction is absolute or graded, and how it is to be applied in specific cases. I spend considerable periods of time repeatedly drawing in particular places. In accordance with Cosgrove, I would have to be classified as an *outsider.* In some locations I see many people engaged in a range of activities. A farmer rounding up sheep would obviously be an *insider.* Would a National Park warden be an *insider?* What about visiting geologists, botanists, etc.? Rock climbers, geography field trip leaders and students? People who hardly stray from a car park and are mainly intent on photographing each other in front of the scenery or the tea stall, one could confidently say, are at the opposite pole from the *insider* farmer, but what of some of these others? Just how useful is this distinction? Moreover, is there never any value in the *outsiders'* possible freshness of vision?

Gendered Landscape

Mitchell links landscape painting with imperialism and the feminisation of nature as a controlling ideology. In discussing John Alexander Gilfillan's painting *Native Council of War (1855)* depicting a Maori gathering, Mitchell remarks on the inclusion of a: "bare-breasted woman, the Renaissance Venus who plays the role of eyecatcher, a titillating bit of soft-core colonial pornography, an emblem of native "nature" opening herself for easy access to the imperial gaze" (Mitchell 2002, 24). Gilfillan is one of many colonial painters who schematically applied the conventions of post-Claudian European landscape painting to exotic regions around the world. The figure of the woman singled out by Mitchell is so diminutive, and so obviously part of the Claudian pastoral landscape tradition being employed, that his claim regarding pornography is totally unconvincing.

Gillian Rose singles out Paul Gauguin's paintings depicting Tahitian women as constituting a prime example of masculinist imperialism: "the visual encoding, of nineteenth-century Western hegemonic masculinist constructions of femininity, sexuality, nature and property are at their most overtly intertwined in the landscapes with figures set in the colonies of Europe and America" (Rose 1993, 94). Landscape as an ideology of power and privilege is, from a feminist point of view, fundamentally gendered in such a way that nature/landscape-as-female is subjected to the dominating male gaze. Rose proposes that: "the same sense of visual power as well as pleasure is at work as the eye traverses both field and flesh: the masculine gaze of knowledge and desire" (Rose 1993, 97–99).

Charles Harrison (1994), in *The Effects of Landscape,* discusses the anthropomorphism of Georgia O'Keeffe's New Mexico landscapes, with particular reference to *Red and Yellow Cliffs* (1940). He reduces the painting to a stereotypical combination of "picturesqueness ... [and] nature as a kind of female body ..., [the latter stereotype being] ideologically normative. ... [The result is] metaphor as cliché" (Harrison 2002, 222).

This criticism that O'Keeffe had produced a stereotypical cliché may be considered more applicable to Alexandre Hogue's *Erosion No.2. Mother Earth Laid Bare* (1938). Hogue's painting shows an area of earth in a field that has the precise form of a recumbent woman. This image of the violation of mother earth by phallic agriculture symbolised by a plough in the foreground, and of the consequent damage of the Great Plains dust bowl, relies on normative associations and values that have turned the metaphor of earth and woman into a stable symbol – a dead metaphor. This enables Hogue to communicate his social and environmental message with great clarity.

O'Keeffe has no such message in *Red and Yellow Cliffs,* the expressive anthropomorphism of O'Keeffe's landscapes is altogether more ambiguous, it serves to establish a sense of empathetic identification with the landscape. I think that this continues to transcend the fate of becoming a dead metaphor, lost to cliché. If not, then the visual articulation of great areas of human experience and sensibility are being systematically closed down by a rapid and inexorable historical process in which creative means bury themselves in dead conventions.

Kitsch Landscape

If the specific conditions of bourgeois hegemony and imperialistic ascendancy productive and definitive of the *landscape idea* no longer exist, then it is plausible to argue that the landscape tradition has degenerated into outmoded conventions as alleged by the Negative Critique. Having become divorced from the contexts that originated and animated it, the tradition is left open to denigration not only for its ideological culpability and epistemological failure, but also for the commonplace inauthenticity that seems to be the major symptom of everything that is wrong with it. If the arguments of the Negative Critique are viable, their living proof could be the kitschness that might be the inevitable and inescapable consequence of continuing to work with these moribund conventions. For Mitchell, one of the main symptoms of the redundancy of the landscape tradition is just this kitsch re-cycling of dead conventions. The intellectual status of any artwork tends to be undermined by genuine popularity. As we have seen with Harrison, this has become a problem for O'Keeffe's work.

Land Art – the Real Thing?

Land Art as a multi-media practice can seem to be more appropriate to the multi-sensual experience of nature. Land Art tends to be presented as a paradigm of creative activity that supersedes outworn traditions. The unpainted landscape can be regarded as nature, while the painted landscape can be regarded as outmoded culture. This view is underpinned by certain strands of environmental philosophy and aesthetics which promote conceptions of aesthetic values taken to be inherent in environmental experience and which can be conceived independently or in explicit differentiation from the aesthetics of art, of painting in particular. The possibility that new dimensions of aesthetic understanding might be opened up by formulating an autonomous aesthetics of environment constitutes an exciting and worthwhile project. The value of this is undermined rather than furthered, if the conception of landscape painting and related aesthetics against which it constructs itself is false.

Andrews (1999) concludes his *Landscape and Western Art* with a chapter on Land Art that explores "the conversion of landscape into land, sometimes as an explicit *act* of restoration" (Andrews 1999, 204). Andrews supports this observation by showing how Land artists have sometimes displaced landscape in art galleries by importing into them the primary, quantifiable, weighable stuff, that the more complex interpretative and perceptual conception of landscape was screening from us. For example, *New York Earthroom* (1977) by Walter De Maria consists "of 250 cubic yards of earth, at a depth of nearly two feet, spread over 3,600 square feet of gallery floor. This is the very stuff of which landscapes are composed and shaped" (Andrews 1999, 203). Land Art or Earth Art can thus be regarded as fulfilling a historical destiny that rounds off the narrative with restorative symmetry. Andrews quotes the Land artist Michael Heizer: "Earth is the material with most potential because it is the original source material" (Andrews 1999, 203). Andrews concludes: "Place is material: material is place" (Andrews 1999, 210). Andrews' concluding construction of "the conversion of landscape into land" is surprisingly simplistic, crudely materialistic and lacking in force of argument. Conveniently, Andrews excludes sustained attention to twentieth century landscape painting from his account altogether and this, in my view, seriously undermines its critical, historical and educational value.

We are sufficiently accustomed to multi-media exhibitions and installations to realise that these maturing forms of artistic realisation and presentation no longer have to rely on affirmations of their practices that rely on critical distancing from painting. The *Artworld,* far from consisting of a single white cube, is a warren of rooms, corridors and spaces that defy mapping as a single contested space. Land Art is a form that generates tired conventions and clichés as well as painting ever has.

Richard Long is described approvingly by Cosgrove for his *ways* of working "directly on and with the land" (Cosgrove 1998, 270), thus exemplifying the insider's experience. A feminist critique of Long challenges much of the critical endorsement of Land Art. Anna Gruetzner Robins quotes Catherine Nash in support of her analysis of Richard Long: "Catherine Nash described male landscape artists 'overtly performing their gender, in neurotic relationship' to the landscape" (Robins 2000, 170).

According to Robins, "Long was the quintessential post-modern male adventurer" (Robins 2000, 168), sustaining the masculinist values of a lone colonial explorer, moreover a politically complacent and "unreconstructed" one (Robins 2000, 171). Robins considers that Long's activities in remote locations amount to the acting out of fantasies that hint at "veiled anger, envy and malice, as Long the male adventurer plays these childish games" (Robins 2000, 171). Robins' deconstruction of Long involves a literal comparison between his relatively comfortable and safe adventures and the hazardous extremes of endurance and fortitude engaged in by actual explorers in the past (Robins 2000, 165). Similarly, Rose (1993), in *Feminism and Geography,* attacks the masculinist nature of geography as a discipline, this masculinism is said to be manifested in the "heroic ethos" of the geography field trip: "Geographers become stronger men by challenging Nature" (Rose 1993, 70).

The Basis for a Positive Critique of Landscape

I have been giving an account of the Negative Critique of Landscape, and I have considered what I regard as some of its shortcomings. A Positive Critique of Landscape would require another chapter. For now, I wish to indicate some elements of such a Positive Critique by confronting a text in environmental philosophy with some paintings. This also leads me to revisit the denigration of pictorial perspective indicated above.

To recapitulate some of the Negative Critique's understanding of landscape and landscape painting, it is interesting to consider some statements from Yi-Fu Tuan's (1974) *Topophilia* in relation to J. M. W. Turner's painting *Light and Colour (Goethe's Theory) – the Morning after the Deluge – Moses writing the Book of Genesis* (1843). This circular composition, replete with scientific and Biblical content, can challenge Tuan's assertions:

> In Europe, some time between 1500 and 1700 A. D., the medieval conception of a vertical cosmos yielded slowly to a new and increasingly secular way of representing the world. The vertical dimension was being displaced by the horizontal; cosmos was giving way to a flat nonrotary segment of nature called landscape. (Tuan 1990, 129)

> When the hydrologic cycle acquired its horizontal dimension it lost its metaphoric power. It became a purely physical process devoid of transcendental and symbolical overtones. (Tuan 1990, 134)

> The ancients believed that movement in nature was disposed toward the circular path. The circle symbolised perfection. The moderns following Newton's revolutionary thought postulated the straight line as the natural path of all moving matter. Cosmos yielded to geography and landscape. (Tuan 1990, 148)

It is not really too fanciful to view the energetic, turbulent spaces of Turner's paintings as being expressive of the thermodynamic and electromagnetic conceptions of matter that were developing in his time, particularly given Turner's personal acquaintance with some of the scientists involved, including Michael Faraday and the scientific writer Mary Somerville. In the paintings, matter becomes incandescent while space is not passively empty but is constituted by vortices of energy. The argument that science and landscape painting wholly diminished the cosmological and metaphorical meanings of nature could hardly be further from the truth.

To Faraday, as to many of his contemporaries in science and philosophy, electricity was a fundamental power of nature:

> ...all the variety of this fair globe may be converted into three kinds of radiant matter, (heat, light and electricity).
> (Faraday as quoted by Gage 1987, 226)

> I assume nothing to be true in any part of space (Whether this be empty as is commonly said, or filled with matter) except forces and the lines in which they are exercised. (Faraday in Spengler 1961 Vol. 1, 416)

A convincing case *for* thinking that Turner made such links in his work is made by James Hamilton (1998) in *Turner and the Scientists.* He suggests that Turner provided an image *for* contemporary speculations and theories regarding electromagnetic forces active in water and ships' machinery in his *Snow Storm – Steam-Boat off a Harbour's Mouth making Signals in Shallow Water, and going by the Lead* ...(1842).

Similarly to Turner, landscape for Paul Nash encompassed cosmological dimensions of man's historical destiny and involvement in the cycles of nature. War paintings, archaeological sites, myth, the seasons, growth and decay were themes of intense personal significance and expression. *Landscape of the Moon's Last Phase* (1944) is one of a series of late paintings dealing with the phases of the moon, and based on views towards Wittenham Clumps from Boars Hill, Oxford. That series achieved a rich consolidation of lifelong themes and practices in Nash's work. The iron-age hill fort emphasised by its planting of beech trees resonates with historical transformation and association. Nash intensified this to create a British landscape archetype that adds much to the Romantic tradition of William Blake and Samuel Palmer.

Anselm Kiefer's mixed-media painting *Emanation* (1984–1986), dominated by a flow of lead from the sky to the sea, seems to affirm Tuan's "vertical dimension" (Tuan 1990, 129). It suggests an alchemical transformatory cycle between Heaven and Earth. While the *vertical dimension* is a recurrent theme in Kiefer's work, railway perspectives constitute another. A dismantled railway line provided the basis for a sequence of photographic works presented in book form: *Siegfried's Difficult Way to Brunhilde* (1977). This creates a fateful sense of time by combining journeying with decay in perspective recession. Collapsing traces of human activity are embodied in the rotting, splintered, shattered railway sleepers. Eventually there are only imprints of vanished sleepers in the stone chippings and weeds that recede in perspective towards a mysterious and long vanished destination. The stark tonality and decisive perspective of the mixed-media painting *Iron Path* (1986) combines the image of railway tracks with real olive branches, redolent of past conflicts and final journeys. Kiefer's use of perspective, often derived from photographs, brings an inexorability to his sense of history, myth and destiny, it is a mixed-media demonstration that pictorial perspective not only remains but continues to develop as a vital component within the field of visual expression and meaning.

Rudolph Arnheim draws attention to the visually dynamic qualities of perspective in pictorial composition. These qualities generate connotations of time and process in the construction of pictorial space:

> ...central perspective locates infinity in a specific direction. This makes space appear as a pointed flow, entering the picture from the near sides and converging toward a mouth at the distance. The result is a transformation of the simultaneity of space into a happening in time – that is, an irreversible sequence of events. The traditional world of being is redefined as a process of happening. In this way central perspective foreshadows and initiates a fundamental development in the western conception of nature. (Arnheim 1966, 287–288)

The idea that perspective lends itself to a visual expression of "an irreversible sequence of events" is particularly haunting in Kiefer's thematisation of railways.

Conclusion

It is my view that, historically, landscape painting has provided a medium for giving form to the relationship of human subjects to their environment in uniquely valuable ways. In landscape, subjectivity and environment reciprocally constitute each other; landscape is a relational form of understanding and creativity. The visual and perspectival connotations of landscape are consistent with this fundamental condition. Stephen Ross develops the function of perspective in the philosophy of Alfred North Whitehead to "argue that being *is* perspectivity" (Ross 1983, 1). The term landscape emphasises relationship in a particular way. This does not make it incompatible with the term environment, it simply refines it. I wish to positively endorse the term landscape and to retrieve it from being tendentiously restricted to ideological meanings.

References

Andrews, M. (1999). *Landscape and Western Art.* Oxford: Oxford University Press.

Arnheim, R. (1966). *Art and Visual Perception.* Berkeley: University of California Press.

Baudrillard, J. (1988). Simulacra and simulations. In M. Postner (Ed.) *Jean Baudrillard: Selected writings,* Cambridge: Polity Press.

Berger, J. (1972). *Ways of Seeing.* London: Penguin Books.

Berger, J. (1980). Millet and the Peasant. In J. Berger, *About Looking.* London: Writers and Readers Publishing Co-operative.

Callicot, J. B. (1998). The land aesthetic. In R. G. Betzler and S. J. Armstrong (Eds.) *Environmental Ethics.* Boston: McGraw-HilI.

Carlson, A. (1998). Aesthetic appreciation and the natural environment. In R. G. Botzler and S. J. Armstrong (Eds.) *Environmental Ethics.* Boston: McGraw-Hill.

Clark, K. (1979). *Landscape into Art.* London: John Murray.

Cosgrove, D. E. (1989). Geography is everywhere: culture and symbolism in human landscapes. In D. Gregory, and R. Walford (Eds.) *Horizons in Human Geography.* London: Macmillan.

Cosgrove, D. E. (1998). *Social Formation and Symbolic Landscape.* Wisconsin: The University of Wisconsin Press.

Crandell, G. (1993). *Nature Pictorialized: "The View" in landscape history.* Baltimore: The Johns Hopkins University Press.

Gage, J. (1987). *J. M. W. Turner 'A Wonderful Range of Mind'.* London: Guild Publishing/Book Club Associates by arrangement with Yale University Press.

Goodman, N. (1976). *Languages of Art.* Indianapolis: Hackett Publishing Company.

Hamilton, J. (1998). *Turner and the Scientists.* London: Tate Gallery Publishing.

Harrison, C. (2002). The effects of landscape. In W. J. T. Mitchell (Ed.) *Landscape and Power.* Chicago: The University of Chicago Press.

Knorr, K. (1991). *Marks* of *Distinction.* London: Thames & Hudson.

Kulka, T. (1996). *Kitsch and Art.* Pennsylvania: The Pennsylvania State University Press.

Mitchell, W. J. T. (1995). Gombrich and the rise of landscape. In A. Bermingham and J. Brewer (Eds.) *The Consumption* of *Culture 1600–1800: Image, object, text.* London: Routledge.

Mitchell, W. J. T. (2002). *Landscape and Power.* Chicago: The University of Chicago Press.

Nochlin, L. (1996). Women, art and power. In N. Bryson, M. A. Holly and K. Moxey (Eds.) *Visual Theory.* Cambridge: Polity Press.

O'Keefe, G. (1976). *Georgia O'Keeffe.* New York: Viking Penguin.

Packer, W. and Angela Flowers plc. (1999). *Contemporary British Landscape.* London: Flowers East.

Pirenne, M. H. (1970). *Optics, Painting and Photography.* Cambridge: University of Cambridge Press.

Robins, A. G. (2000). "Ain't Going Nowhere". Richard Long: Global explorer. In S. Adams and A. G. Robins (Eds.) *Gendering Landscape Art.* Manchester: Manchester University Press.

Rose, G. (1993). *Feminism and Geography.* Cambridge: Polity Press.

Roskill, M. (Ed.) (1977). *The Letters* of *Van Gogh.* Glasgow: Fontana/Collins.

Ross, S. T. (1983). *Perspective in Whitehead's Metaphysics.* Albany: State University of New York Press.

Schama, S. (1996). *Landscape and Memory.* London: Fontana Press.

Spengler O. (1961). *The Decline of the West.* 2 Vols. New York: Alfred A. Knopf

Taylor, J. (1994). *A Dream* of *England.* Manchester: Manchester University Press.

Tilley, C. (1994). *A Phenomenology* of *Landscape.* Oxford: Berg.
Tuan, Y. (1990). *Topophilia: A study* of *environmental perception, attitudes, and values.* New York: Columbia University Press.

10. THREATENED ENVIRONMENTS, ATROPHYING CULTURES, LACKLUSTRE POLICIES

COLIN H. WILLIAMS

Introduction

Interpreting the relationship between culture and environment has been one of the preoccupations of human thought for time immemorial. More recently, concern about threats to the natural environment, in both general and in specific locales, has been accompanied by additional considerations related to the survival of endangered peoples and cultures, to the overt discrimination against specific groups within multicultural societies and to the relative powerlessness of sections within society. Typically, such matters involve issues of native rights, land despoliation and pollution, sustainable development, neo-colonialism and globalisation, with the environment being very differently presented and represented by the various parties involved. The most extreme and urgent cases of concern are located in regions of Latin America, Central Africa, South-East Asia and Australasia, where evidence of global warming and the depletion of biodiversity have led to calls for global governance and the assertion of a trans-national liability for environmental damage. However, the increased recognition that the cultural-environmental nexus is a global continuum has led to a greater consciousness that similar if less disastrous cases of degradation and harm are also found in the more mature industrial societies. In these, evidence of environmental despoliation, cultural attrition and denial of group rights bespeak a common global condition. Both the United Kingdom and Canada provide examples comparable with those found in countries of the less developed world. The similarities extend further in the way that the relationship between specific dependent regions and the central state apparatus influences what is possible and determines the differentiated treatment of citizens within their respective territories.

There is usually an integral relationship between the local cultural and natural environment within which a minority culture group is embedded within the wider state. The attitudes, perceptions and values of such a minority group will be fundamentally different to those of the majority with dominance over the wider state area. A wide variety of interests will be evident on both sides, each wishing to present and

G. Humphrys and M. Williams (eds.), Presenting and Representing Environments, 153-166.

represent the environment in different ways to achieve very different and often competing outcomes. In this chapter this is revealed in the way the relationship between ecological movements and the new localism is resonant with meanings about the tensions between the 'global' and the 'local,' between individual acts of socio-economic development and the communal defence of threatened environments, and particularly between trans-national structural integration and ethnic-regional struggles for relative autonomy.

Basic Issues

Notions of community defence have been commonplace since at least the early mediaeval period. But it is only more recently that a self-conscious concern with issues of social fragmentation and community decline has given rise to the need for active revitalisation and empowerment of threatened communities. Community empowerment is an attractive notion but one which citizens find increasingly difficult to realise because of the complexity, scale and pace of socio-economic change. Many assert that community ownership of the processes of social reproduction is made more difficult within bilingual, multilingual or language border land societies, because of the structural tension inherent in a situation of competitive language contact. I argue that it is possible to counter such conventional arguments as long as the structural preconditions of community involvement are in place to allow the trigger of interventionist planning to instil a fresh sense of ownership of the community's integrity and destiny.

Today, many societies in Europe, such as those in Brittany, Estonia, Euskadi, Slovenia and Wales, are undergoing a process of territorial restructuring and socio-political transformation as they adjust to a new European political environment. Geo-strategic and economic questions are inherent in this process of adjustment, but so also are basic issues related to more universal questions and tensions regarding the role of humans as agents in natural environmental change. These tensions derive from our attitude to a fundamental question of moral philosophy: Are humans masters over nature? In previous work I have sought to answer this question by reference to four distinct human-nature positions derived from Jewish, Graeco-Roman, Christian, Modern and other views of nature. The positions are respectively that: nature is neutral and without purpose; nature is a divine creation in the service of man and devoted to the Glory of God; nature as a divine creation is good for its own sake; nature has a meaning and a purpose and reflective man must ascertain that purpose and accommodate it (Williams 1991; 1999). Current manifestations of ecological thought can be traced to each of these separate positions, but are most commonly associated with the fourth position whereby powerful ideas, such as the Gaia thesis and the 'deep green' concern with earth survival and co-existence, are connected to earlier traditions of humanistic thought.

Intellectual revisionism suggests that what passes for contemporary environmental thought can be traced back to earlier epochs in the long inter-relationship between human speculation and natural conditions. What may be seen as new about environmentalism is its global concern and sense of urgency. But we need only remind ourselves of the universal appeal of major world religions, with their imminent need for 'salvation' from sin, destitution and the apocalypse to come, to realise that both globalism and imminent disaster have been central planks of earlier, more persistent, ideologies. Neither can we accept the revelation that environmentalists have discovered a new truth about humanity. Drawing on several earlier traditions, the Humanists of the Enlightenment could support Hume's dictum that "In all nations and ages, human nature remains still the same" (Hume 1975, 34). Interpreting the universal history of mankind led enlightenment thinkers to argue that *reason* was the same for all. Yet, Solomon (1980) suggests that humanism was, in fact, "a new discovery of sorts, a grand abstraction without geographical or anthropological limits". For, "in the name of Enlightenment humanism, the middle-class philosophers could declare their interests to be universal interests, and their interests to be the dictates of reason. And since their reason was the very heart of human nature, enlightened self-interest was never merely personal but in the interests of everyone. To emphasise the universality of such interests, therefore, they could be turned into universal laws and it is very significant that the enlightenment philosophers insisted on 'government of laws, not men'"(Solomon 1980, 29–30). What is new is the politicisation of issues such as acid rain, marine pollution, land degradation, deforestation and desertification, and the realisation that, because of the global inter-connectedness of environmental destruction, the effects are never merely local. Economic growth is no longer viewed as the apex of civility and progress, and is increasingly being blamed for much of the world's problems. It is the scale and intensity of world economic growth that poses the ultimate threat to the biosphere.

> How else can one explain that more destruction has been wrought to the fragile fabric of the biosphere during the last 40 years, since global development has really got under way, than during the preceding two or three million years of the human experience on this planet?
>
> (Goldsmith 1988, 2)

The question remains whether we can learn anything from former ideologies and mass movements for social change? I believe that there is a particular merit in focusing on the comparisons that can be made between environmentalism and other *-isms*, particularly communalism and nationalism.

Environmentalism and Nationalism

A key comparison of both ideologies is that environmental and regional-nationalist movements focus on the state as the central agency of 'structured oppression' and the principal upholder of the military-industrial complex. Galtung (1986) has summarised the opposing characteristics of establishment and ecological values and movements and argued that many autonomist movements share several of the characteristics attributable to green policies. Thus, for example, many nationalist movements in post-war Europe have adopted a non-violent form of resistance to the incursions of central state agencies. There has often been an acute sense of divine destiny in the mobilisation of minority cultures, whether in Catholic Catalonia and Euskadi (Conversi 1997), Ireland (Goldring 1993) or Nonconformist Wales (Llywelyn 1999; Morgan 2001). Equally revealing has been the concern with localism, with self-reliance and with non-alignment, in order to distance supporters from the hegemony of the superpowers and strong states. In their presentations and representations at root both environmental and minority autonomist rights movements seek to place the individual within a wider communal framework, stressing inter-dependence and a shared destiny. Feelings of belonging, of shared responsibility and of rootedness figure prominently, in contrast to the possessive individualism of post-Fordist culture which stresses mobility, individual advancement and regime of flexible accumulation.

A similar litany can be constructed for European ethno-regionalist movements which are prompted by fears of loss of identity, exploitation of the cultural and physical resources of the nation, resistance to external intervention and control, a reinterpretation of the nation's plight in terms of the leadership's mission-destiny view of their own transcendent existence, and the ultimate securing of freedom in a world of free and equal nations. This nationalist iconography has raised a triad of political truths, sovereignty, autarky and cultural integrity that the Welsh philosopher J. R. Jones described as *cydymdreiddiad iaith a thir*, the mutual inter-penetration of land and language (Jones 1970; Williams 1988).

Ethno-regionalists invariably insist on the defence of their unique homeland, and of the protection of their valued environment, an expression of place-centred politics, whereby the politics of collective defence over land and territory comes to dominate the relationship between a minority movement and its incursive, hegemonic state power. More recently ecologists have taken up this concern with valued environments and in some cases entered into a common alliance with nationalists to protect threatened spaces. This seems set to become a major issue of the twenty-first century where conflict over the control of space and resources will quicken our awareness of the commonality of defensive movements, even if their ideological rationale remains distinctly separate.

A more persistent tendency is for local communities to resist population transfer and demographic changes that adversely affect the majoritarian position of co-nationals. Thus, resistance to outsiders, to immigrants, to settlers, to colonisers as agents of the hegemonic state and its associated culture, is as widespread in rural Wales as it is in urban Euskadi. This is related to a fear of loss of local dominance and influence and to ineluctable cultural attrition. It spawns campaigns to save local communities, sacred sites and valued land from external development. Several campaigns in Wales, such as at Rhandirmwyn, Tân yn Llyn, *Nid yw Cymru ar Werth* (Wales is not for sale), *Dal dy Dir* (Hold your land), all demonstrate resistance to changing the socio-economic infrastructure to suit external interests. Similar examples are to be found in Brittany at Plogouf, in opposition to nuclear power or second home/tourist development, and in Scotland in relation to nuclear installations and defence establishments. Most often, dangerous and obnoxious industries, such as nuclear power plants, oil terminals, and defence establishments, are situated within the minority's territory, for the communities are relatively powerless to stop such exploitation of their land and social fabric. Indeed, in some cases the minorities are keen to welcome these installations since they contribute to economic survival. However, the exploitation from elsewhere of the natural and human resources found in the territory of minorities can induce a dependency situation that is reactive rather than purposive, defensive rather than self-confident, outer-directed rather than inner-directed. Structural tension can lead to conflict and sustained violence, as in Euskadi, Ulster and Corsica. Attempts to counter such violence, through various stages of regional autonomy, power sharing and state re-structuring, do not appear to be wholly successful. Minority nationalist intelligentsias have sought to influence group learning by stressing in their representations those aspects of a real or mythical history that would explain contemporary inequalities in the light of significant acts of oppression. They have constructed alternative versions of their historical reality rooted in a different trajectory and communicated through a separate language (Williams 1988).

Centralists express their virulent opposition to formal education in the minority's mother tongue precisely because they fear they will be unable to interpret and control the messages circulated through the minority languages. Frequently, the charge has been made in parts of Europe that minority schools (for language, ethnic and religious minorities) are the breeding ground for a new generation of dissenting groups. Throughout Western Europe and North America there are new demands for centrally controlled educational curricula rather than a deepening of differentiated regional educational varieties that may, in part, perpetuate group divisions and sectarian animosities. Thus, contemporary nationalism, like environmentalism, seeks to be holistic in thought and deed. The universal desideratum is an ideal nation-state, defined by selected socio-cultural criteria, including language, in an international community of free and equal states. Its local manifestation may be at any one of the varying stages in a continuum from full statehood through to a non-state nation in the making, but whatever its exact position it is always possible to relate the local to the global blueprint and back again. So it is with environmentalism. The unique character of the local environment is only given purchase by the general context of the global milieu. It is given urgency by the near-cataclysmic refraction of fundamental global issues, of the greenhouse effect, the destruction of the tropical rain forest, the degradation of soil and sustenance, and the disappearance of species and habitat. Intrinsic to this coupling of the immediate and local to the evolutionary and global is a sense of shared involvement and responsibility, summed up so cogently in the slogan 'Our common home. Our common future'.

In several societies there is active co-operation between ethnic-regionalist (nationalist) and ecological (green) movements. In the expanded European Union, two major developments have served to bring both types of movement to centre stage. First, the more fundamental event is the increased recognition of the cogency of the ecological argument at all levels. At the macro-level, there is the central concern of the politics of global survival, with all its principled implications for socio-economic existence under the nuclear threat/umbrella. At lower levels in the hierarchy, there is a concern about achieving sustainable development, about limited growth, about limited local control over immediate environments and about limited opportunities for communal expression. Local politics is increasingly the politics of resource allocation and redistribution, where questions of appropriate economic development, raising the social quality of life and ecological regeneration, particularly in sprawling metropolitan areas, are the norm.

There is also a challenge from the apparently inexorable move towards an enlarged and functionally integrated European Union (EU). This is likely to force a re-alignment of the role of majority and minority national groups in Europe. Some regional and national elites, whilst maintaining their negotiations and working relationships with central core agencies, may also seek to entreat directly with corresponding regional-level decision-makers and construct a reformulated, quasi-federal Europe, at least in functional if not in juridical terms. Currently there is a plethora of bi-lateral partnerships within selected regions of EU states, with Catalonia, Scotland and Bavaria being prominent. As the process of adjustment to new political realities unfolds, there is likely to be an increased pressure for new approaches to tackle old problems. Though by no means certain of success there is a new cogency to old decentralist arguments for a regional scale governmental structure throughout Europe. In seeking to bring power back to the people at a recognisable scale, some nationalists, liberals and ecologists are in favour of creating a Europe of the Regions (Williams 1989). This idea is likely to find most support in the relatively underdeveloped regions of Europe and where an ethnically differentiated minority is striving for greater autonomy in order to preserve and protect its threatened identity in a hostile or disadvantageous environment. It is also likely to gain wider acceptance as the meso-scale of government achieves more prominence in an increasingly integrated European space. One of the key policy areas where such initiatives are played out is the environment. Here, the more holistic interpretations, transposing the writings of influential scientists such as Lovelock (1979; 1988) and his Gaia thesis into the realm of social and political affairs, have gained currency. This new 'self-consciousness' has been mobilised both by agencies of central government and by the leadership of those communities who feel in some way threatened by the developments of centralised bureaucracy and the economic imperative of resource extraction and exploitation by commercial interests.

Historically land has been the single most significant element of the environment to be fought over by those competing interest groups wherein the conflict between the local and the universal, the indigenous and the exogenous groups, have been mediated. But, as argued elsewhere, territory is not merely contextual, it is not only a scene for the playing out of socio-economic factors, it is itself a significant source of symbolic and resource power (Williams 1988). Territorial possession and control are often deemed vital to the very survival of specific cultural groups. Consequently, the rest of this chapter will address this issue in relation to the contestation of space and community empowerment in Wales.

The Communal Defence of the Environment

Today, our conception both of a 'threat' to and 'defence' of communal interests has widened. Though central, territory is no longer synonymous with environment. Communal action is often non-institutionalised; it cannot be captured within the simple description of party politics. Many activists have plural constituencies and interest groups that they can mobilise in defence of what they define as 'their environment'; others have none, hence the difficulty in generalising about such instances. Given that the Welsh ethno-linguistic minority is also relatively under-developed in economic and political terms, it becomes evident that questions of language, culture and identity are not supplementary to more routinised socio-economic concerns but can themselves constitute the very essence of that relationship of the minority with the state, in whose name the majority exercise power and control. As such, expressions of group identity are rooted in the environment and derive both material and spiritual sustenance from the immediate locale and it is not surprising that any threat to the environment is interpreted as a simultaneous threat to both the minority culture and group survival. Place and territory are central to this process of control and development by the minority, and their appropriation has a long history associated with the development of the European state structure (Williams 1989).

As a consequence of industrialisation and modernisation, Wales was more fully integrated with the British economy and the expanding world-economy. Simultaneous developments within English regions were interpreted as being advantageous and conducive to national growth, save for the class-based exploitation of workers and the transformation of specific landscapes by mineral extraction, railway construction and factory expansion. However, precisely because of national differentiation and external control, equivalent developments in Wales were considered by some to be damaging to Welsh life, exploitative and counter-productive. Nationalists and some socialist activists were the most prone to interpret Welsh economic change as deriving from colonial development by England. Whilst such an emphasis was crude in analytical terms, it generated a remarkable appeal and facilitated the development of a number of political movements supporting Welsh autonomy: notably Plaid Cymru, The Welsh Nationalist Party, formed in 1925 (Williams 1988). The environmental transformation of Wales after 1850 derived from a series of complex and often mutually contradictory forces. However, the most significant transformation was the extraction of mineral resources and the construction of a Welsh urban hierarchy accompanying the rapid industrialisation. The slate, coal, iron and steel industries were to make Wales a pivotal resource base for the development of the British Empire. In industrial Wales the effects were dramatically imprinted both on the population and on the landscape.

> Its social structure, settlement patterns, population composition and incipient radicalism all testified to the wholesale ravages wrought by the badgering of its coal-rich seams. The marginality of Wales's pre-industrial existence was swamped by a frenzy of commercial and social

> activity, the like of which had not been seen in Europe. Massive dislocation, an immigration rate second only to that of the United States of America, and the reconstruction of local issues and concerns as British issues and imperial concerns, transformed the insular periphery into an active participant in Britain's quest for global hegemony.
>
> (Williams 1989, 229)

A second theme has been a struggle for control of the environment, particularly the acquisition of farmland for large-scale infra-structural developments. A poignant episode was the symbolic attack on a British defence establishment, the Royal Air Force (RAF) bombing school at Penyberth in North Wales. In September 1935 three leading nationalist intellectuals, Lewis Valentine, Saunders Lewis and D.J. Williams, set fire to the sheds of the bombing school and afterwards gave themselves up to the local police. At their trial in Caernarfon, the defence argued that, by seeking to defend one of Europe's oldest and most threatened cultures, the plaintiffs were exercising their convictions based upon a higher moral order than state ordinances. This alleged intrusion into a predominantly Welsh-speaking region of the RAF School, perceived as a symbol of British imperialism and militarism, was recognised by some of the jury who failed to agree on a verdict. The case was transferred to London where the three were found guilty and imprisoned on 19 January 1927 only to be released on 27 August 1927, to a heroes' welcome (Davies 1993). In strict terms this was a minor incident, but nationalist historiography has raised the defendants to the status of Nationalist martyrs, willing to defy the state in the name of securing the cultural integrity of the local community. In post-war Wales, there were further demonstrations when war-time training grounds were not returned to private ownership but were retained by the Ministry of Defence. Parallel decisions in England evoked early Campaign for Nuclear Disarmament (CND) opposition, but lacked the nationalist dimension of outrage and resistance so characteristic of Wales.

In civil affairs, also, some Welsh communities struggled to preserve their rural environments and farming. Post-war reconstruction and the growing modernisation of industry in England necessitated increased investment in the water-supply service. In 1955, Liverpool Corporation sought permission to drown the valley of Tryweryn in mid-Wales to create another reservoir. Penllyn was sparsely populated and ideally suited from a resource perspective. But it was also an area rich in cultural heritage and epitomised all the virtues of rural Welsh-speaking areas that activists sought to preserve as a context for a distinct Welsh 'way of life' (*buchedd*). Despite an unprecedented campaign of Welsh solidarity, with twenty-seven of the twenty-eight Welsh Members of Parliament (MPs) present voting against the passage of the Bill through Parliament, a huge majority of British MPs voted in its favour. Organised Welsh public opinion recognised that, despite having all the elected and representative bodies supporting the campaign, Welsh interests were a permanent 'minority issue' in a democratic British union. '*Cofia Tryweryn*' (Remember Tryweryn) became a rallying cry for some young people, and came to symbolise Welsh dependency on the British state prerogative and power.

A further instance of state interests seeking to dominate communal needs was the decision to site early nuclear-powered electricity generating stations at Trawsfynydd and Wylfa, both in the Welsh language heartland of north-west Wales. This was a decision that an earlier generation of local politicians and planners accepted as providing much needed employment in rural Wales. However, in the 1980s, when the Central Electricity Generating Board sought to build a second nuclear-powered generating set at Wylfa on Anglesey, local and national environmental and cultural-political movements combined to form PAWB (People against Wylfa B). PAWB provided a detailed set of alternative proposals and mobilised support. After a well-conceived and highly articulate campaign, PAWB celebrated a partial victory in 1989 when the proposed Wylfa expansion was postponed indefinitely.

A final theme is that of external private capital investment in selected parts of Wales. At the macro-level, large petro-chemical complexes have been built, as at the Milford Haven and Amlwch anchorages. These are characterised as potentially damaging installations if chemical explosions or oil tanker spillages occur, while the Milford Haven installation has been built within the confines of a National Park established to conserve a coastal environment. At the micro-level, there has been rapid expansion of tourism and second-home ownership in Wales. The latter has been the subject of increased controversy, with the media, perhaps understandably, focussing on the two hundred or so arson attacks on second homes, mainly in north and west Wales, rather than the reasons for the underlying deep-seated grievances. The arguments are familiar and need not be repeated here except to instance that cultural activists have realised that the key to ethno-linguistic maintenance is adequate employment, suitable housing, a reasonable infrastructure and sustainable development, a realisation reflected in the slogan '*Nid yw Cymru ar Werth*' (Wales is not for sale).

Post-Devolution Policies and Discourses

The Wales Assembly Government (WAG) has outlined its position on the future of the Welsh language in the strategic document *Iaith Pawb* (Wales Assembly Government 2003). Following a comprehensive review of the state of the Welsh language undertaken during 2002, the government has committed itself to achieving five goals:

1. By 2011 to increase the proportion of Welsh speakers by 5 percentage points from the 2001 Census baseline.
2. Arrest the decline in heartland communities, especially those with around 70 per cent or more Welsh speakers.
3. Increase the proportion of children in pre-school Welsh education.
4. Increase the proportion of families where Welsh is the principal language.
5. Mainstream Welsh-medium services.

An Action Plan was produced which set out a number of key targets with policy options for achieving the aims, and there is a national strategy which includes a number of elements designed to strengthen the use of the Welsh language. These include the operation of the principle of Welsh language equality with English and relating its bilingual practices to other levels of government, institutions and to civil society.

The implications of *Iaith Pawb* for policy development in Wales in several sectors have been analysed elsewhere (Williams 2004). The rest of this chapter will concentrate on the relationship between community and environment by assessing to what extent government proposals are likely to tackle the atrophying of Welsh speaking communities.

Iaith Pawb refers to, but does not fully address, the issues of in-migration, community development and the housing market. These factors have been perennial sources of concern in modern Wales, and political parties and socio-political movements such as Adfer, Cymdeithas yr Iaith and Cymuned have been active in seeking to ameliorate them. Cymuned (Community), a group established in the summer of 2001, is the most recent voice to represent the protection of the Welsh language within its historically significant communities, largely in the north and west of Wales. Comprised of local councillors, intellectuals, and community activists, Cymuned bridges a gap between the established political parties and the more radical, if well established, direct action stance of Cymdeithas yr Iaith Gymraeg (Welsh Language Society). Cymuned is a community action group with a linguistic cutting edge. It seeks to present the issue of linguistic and socio-economic rights in Wales as a national variant of an international trend, namely the atrophying of historically well-grounded communities by state policies and the forces of globalisation. Its ideological stance is rooted in communitarian nationalism but it has a specific applied focus in that housing and the integration of non-Welsh speakers are the twin foci of campaigns.

In the presentation of its case Cymuned clearly identifies the cultural environment it is seeking to sustain. It argues that traditional Welsh speaking communities in rural Wales face the problem that "a very vibrant and rich minority culture is being threatened by a combination of socio-economic factors" (Cymuned web site 2004). It identifies as a key element the need for local people (English and Welsh speaking) to have the right to buy houses in their own communities in the face of in-migration by English speakers who can pay higher prices. It is also concerned about the influx of individuals who refuse to learn the Welsh language and are therefore unable to contribute to what is represented by language advocates as an indigenous Welsh speaking cultural and social life. Cymuned has also fashioned a coherent set of proposals that it wishes to see discussed in what passes for civil society in Wales and beyond. It has presented these in evidence to the United Nations, the Welsh National Assembly and to the Richards Commission which was established to look at the development of the powers of the Welsh Assembly Government. Basic to its argument is its belief that the definition of acceptable bilingualism and of equality of language status that is promulgated by government in Wales is flawed. "This is not a mechanism for securing the continuance of the Welsh language as a living social medium: the end result of the type of bilingualism conceived of by the Assembly is anglicisation" (Cymuned 2003, 5). A sample of the constitutional proposals made by Cymuned to the Richards Commission in 2003 indicates the essence of its position. Cymuned called for the setting up of a Welsh Heartlands Authority by the Welsh Assembly Government (WAG) to take over all executive, representative and advisory responsibilities at local government level, in the fields of planning, housing, economic development, education and communications in specific areas. Permanent residents of these areas should be guaranteed the right and ability to conduct all aspects of their daily lives entirely through the medium of Welsh. Public, private and voluntary business in these enclaves would be conducted solely through the medium of Welsh, with the provision of full translation and interpreting services into and out of English as and when required by non-Welsh speaking people. Non-Welsh speaking residents would be enabled to speak the Welsh language. The strength of Cymuned's case is that unless urgent ameliorative action is taken across a whole range of socio-economic activities, many Welsh-speaking communities will continue to atrophy. Specification of the language-related rights of both individuals and communities in Wales would also bring to the fore the current inadequacies of statutory provision of services and working practices. However, the weakness is that establishing a Welsh Heartland Authority will not achieve the desired results. Further, the expectation that most individuals and certainly most employees within the Heartland-defined economy would or should be Welsh speaking is unlikely to be fulfilled (Williams 2002). Socio-economic networks and occupational mobility are not congruent with local authority boundaries, save within selected aspects of public administration and education. Despite these flaws, there is a case for arguing that several of the practical suggestions of Cymuned should be implemented in a limited, experimental fashion so as to identify those that are likely to succeed (for details, see Williams 2004).

Waiting for a Heartland Authority to be established is counter-productive. This approach was tried in the 1960s and 1970s when Betts (1976) and Williams (1977/8) tackled the issue of institutionalising *Y Fro Gymraeg* as a local government and planning authority. It could be argued that the need is now more acute. However the same political imperatives remain. Governing political parties of whatever complexion are highly unlikely to institute territorial reforms that will deliver a near-permanent power base to their political opponents. It would be far better to implement several functional proposals in a more limited way to tackle the current malaise through a combination of inter-agency co-operation and government-directed policy. More realisable aims have been presented to strengthen government policy, including: a statutory status for Welsh; a national language planning framework; increased finances for language planning; an adequate framework for housing and planning regulation; an extension of the Welsh Language Act 1993 to the private and voluntary sectors; greater provision of Welsh medium education; normalising Welsh in 'traditionally Welsh speaking' communities; the relocation of Assembly offices beyond Cardiff and the South Wales Valleys; and the need to ensure that future Censuses provide more useful relevant data.

Probable Government Response

Though voiced by Cymuned and Cymdeithas yr Iaith, their proposals are the substance of much official agency discussion with the Welsh Assembly Government. This recognition of official agency discussion is important, for many reforms in Wales are undertaken by an alliance of informed networks, often cross-party or non political in nature. They are rarely undertaken as a reaction to direct pressure or campaigning only. Interim schemes already implemented include 'Affordable housing', the Homebuy scheme-with grants of 50 per cent in rural areas, the Social Housing Grant programme (£56.4m in 2003/4), the operation of the Housing Needs Assessment scheme, research on second and holiday homes, together with studies on the planning implications of changing housing ownership legislation. WAG is also advancing a broader review of the town and country planning process and its effects on the Welsh language. *Iaith Pawb* references this in relation to revisions of WAG guidance to the planning profession, to housing development guidance, and to conducting linguistic impact assessments (LIAs) currently being devised in a joint-venture which includes thirteen local authorities, two National Parks, the Welsh Language Board and academic specialists.

In recognising that both community fragmentation and the failure to transmit Welsh within the family are serious issues in the minority cultural environment, *Iaith Pawb* has sought to build on existing remedial solutions. The successful Twf – Family Language Transfer – programme is to be expanded, as is the enhancement of operational bilingualism within The National Health Service (NHS) Wales and the extension of *Iaith Gwaith*, a scheme to promote Welsh in the workplace. *Iaith Pawb* also advocates the creation of a national database of standardised terms, lexicographical and machine translation provision, a rigorous translation strategy and a plethora of information technology advances to encourage bilingual/multilingual communication. In truth, many of these initiatives are but the repackaging of existing programmes, albeit with a greater insistence on mainstreaming Welsh. Thus, bilingualism is to be encouraged within the economy through Finance Wales plc; the Community Loan Fund; the Community Development Financial Institutions and integrated more fully within the WAG *Wales for Innovation* strategy. Welsh culture is to be marketed through the 'Wales' Tourism Product, various Cultural Tourism Strategies, and through the Destination Management System. Community development receives attention through schemes such as the Communities First programme, the Welsh Development Agency's Regeneration Framework, Farming Connect, Retail Outlets, the Post Office Development Fund, the Wales Rural Observatory and, of course, the invaluable work of the *Mentrau Iaith* and the Leader related project within a broader Rural Community Action scheme and Rural Development Programme.

Mainstreaming bilingualism is also the keynote for reforms within Welsh public administration. Fifteen instruments or avenues have been identified to achieve this end. They include: the issuing of guidelines or circulars; NHS instructions; policy agreements with local government; setting standards; inspection, review and audit of programmes; regulations and statutory guidance; approving schemes; performance management systems; core grant allocation and general funding agreements with partners; contracting arrangements, especially with third parties, are expected to specify in tender documents the nature of service provision and pertinent linguistic considerations; partnership agreements are expected to operate in accordance with the body's Language Scheme and *Iaith Pawb*'s recommendations; incorporating Welsh language considerations into the core of an equal opportunities agenda; initiating research into how Welsh speakers may be better served; consciously developing information technology systems which facilitate rather than frustrate the use of Welsh; and general organisational issues which promote bilingualism.

For many the sticking-point in the overall strategy is the lack of political leadership on the relationship between the language and the economy. After all, much of the real concern with atrophying communities and others in Wales is not language-driven, rather it reflects deep economic malaise. Despite recent promising initiatives undertaken by *Menter a Busnes*, *Antur Teifi*, the Welsh Development Agency (WDA) and the partners represented within the Language Economy Discussion Group, much remains to be done to promote the use of Welsh in the economy. Future efforts are likely to focus on: securing the establishment and interrogation of national data sets regarding the language-economy nexus; identifying conditions which enable Welsh to be used as a language of work; attractive promotional campaigns in many promising contexts; advising employees and consumers of their language rights under a combined Equal Opportunities/Welsh Language Act/*Iaith Pawb* umbrella of duties and expectations; socialising external companies as to the best bilingual practice available when selling or distributing their products in Wales; and a further professionalisation of the commercial and economic development aspects of language planning in Wales.

Conclusion

Many of the current policies discussed above can be considered sound and well intentioned, but their application is lacklustre because of a lack of conviction at both national and local level. Issues of community vitality, bilingual socialisation and environmental sustainability are treated as if they operated within separate microcosms with very little inter-connection. Despite all the rhetoric that trumpets holistic thinking and joined-up planning, Wales is far from achieving robust policies that are conducive to sustainable development for both people and the environment. Presenting and representing environments is, of course, a continuation of power politics by other means. Vested interest groups and contesting political parties are unlikely to implement policies that are conducive to the long-term sustainability of Welsh health society in this context. And yet, this is precisely what the more far-reaching visionary reforms in the past which benefit current Welsh society did. Cynics would encourage the belief that nothing will really change and it is futile to hope for the future Wales deserves. The condition of the Welsh language cultural environment in particular is too fragile and the community base too fragmented to do other than fight for improvement using democratic means. To do other is to misrepresent the environment that is cherished and present a victory to the cynic, the fatalist and the architects of sterile uniformity.

References

Betts, C. (1976). *Culture in Crisis.* The Wirrall: The Ffynnon Press.

Conversi, D. (1997). *The Basques, the Catalans and Spain.* London: Hirst.

Cymuned (2003). *A Submission by Cymuned to the Commission on the Powers and Electoral Arrangements of the National Assembly for Wales.* Aberystwyth: Cymuned.

Cymuned (2004). News and releases from the web site. www.cymuned.org/news (accessed October 2004).

Davies, J. (1993). *A History of Wales.* London: John Allen The Penguin Press.

Galtung, J. (1986). The green movement: A socio-historical exploration, *International Sociology,* 1, 75–90.

Goldring, M. (1993). *Pleasant the Scholar's Life: Irish intellectuals and the construction of the nation state.* London: Serif.

Goldsmith, E. (1988). Foreword. In E. Goldsmith and N. Hildyard (Eds.) *The Earth Report.* London: Mitchell Beazley.

Goldsmith, E. (1989). Towards a biosphere ethic, *The Ecologist,* 19, 68–75.

Hume, D. (1975). *Enquiries Concerning Human Understanding and Principles of Morals by David Hume: Reprinted from the posthumous edition of 1777.* Edited by L. A. Selby-Bigge. Oxford: Clarendon Press.

Jones, J. E. (1970). *Tros Gymru.* Swansea: Tw John Penry.

Lovelock, J. (1979). *Gaia: A new look at life on Earth.* Oxford: Oxford University Press.

Lovelock, J. (1988). Man and Gaia. In E. Goldsmith and N. Hildyard (Eds.) *The Earth Report.* London: Mitchell Beazley.

Llywelyn, D. (1999). *Sacred Place, Chosen People: Land and national identity in Welsh spirituality.* Cardiff: University of Wales Press.

Morgan, D. D. (2001). *Cedyrn Canrif. Crefydd a chymdeithas yng Nghymru'r ugeinfed ganrif.* Caerdydd: Gwasg Prifysgol Cymru.

Solomon, R. (1980). *History and Human Nature.* Brighton: Harvester Press.

Wales Assembly Government. (2003) *Iaith Pawb. A National Action Plan for a Bilingual Wales.* www.wales.gov.uk/subiculture/content/iaethpawb.pdf (accessed October 2004).

Williams, C. H. (1977/8). Cynllunio Ar Gyfer Yr Iaith Yng Nghymru a'r Sefyllfa Mewn Gwledydd Eraill, Rhan 1, *Barn,* Rhif 179; Rhan 2, *Barn,* Rhif. 180.

Williams, C. H. (1988). Minority nationalist historiography. In R. J. Johnston, D. Knight and E. Kofman (Eds.) *Nationalism, Self-determination and Political Geography.* London: Croom Helm.

Williams, C. H (1989). The question of national congruence. In R. J. Johnston and P. J. Taylor (Eds.) *A World in Crisis?* (2nd edn.). Oxford: Blackwell.

Williams, C. H. (1990). The Anglicization of Wales. In N. Coupland (Ed.) *English in Wales.* Clevedon, Avon: Multilingual Matters.

Williams, C. H. (1991). Language planning and social change: ecological speculations. In D. F. Marshall (Ed.) *Language Planning: Focusschrift in honor of Joshua A. Fishman.* Amsterdam: John Benjamins.

Williams, C. H. (1999). The communal defence of threatened environments and identities, *Geografski Vestnik,* 71, 105–120.

Williams, C. H. (2002). *Adfer yr Iaith: Cynllunio, Yr economi a thiriogaeth.* Papur Trafod, Plaid Cymru, Caerdydd, 2002/ '*The Revival of the Welsh Language: Planning, economy and territory*', Commissioned Policy Advice for Plaid Cymru, Policy Discussion Paper. Cardiff: Plaid Cymru.

Williams, C. H. (2004). *Iaith Pawb:* The doctrine of plenary inclusion. *Contemporary Wales,* 17, 1–34.

11. SUSTAINING ARCTIC VISIONS, VALUES AND ECOSYSTEMS: WRITING INUIT IDENTITY, READING INUIT ART IN CAPE DORSET, NUNAVUT

NANCY C. DOUBLEDAY

Relative to southern Canada, the Arctic has been and continues to be perceived as being a region of a different order: whether as 'frontier' or 'homeland', as resource storehouse or park reserve. Described variously as a land of contrasts, of light and dark, of extremes, clearly the idea of the Arctic is represented in many ways and serves diverse interests. In turn we can employ this dualism in a variety of ways to achieve many intellectual ends. Here it is seen as a tool of perspective, allowing us to interrogate points of view, particularly regarding the interplay of environment and culture, focusing on the Arctic but reflecting critically in the end on our own assumptions.

Arguably the fundamental dualism, that of frontier and homeland, reflects projected desires from multiple perspectives, but it can also be seen to represent a long-standing conflict within Western traditions of nature and environmental protection, in which humans are positioned apart from the environment. Ultimately, even these categories can be claimed to be 'contested' just as the nature of the Arctic is claimed to be, an assertion which is to be explored here, through reflections on the presentations and representations of the Arctic in the indigenous Inuit art of the region.

While Canada has a deep culture of natural landscape appreciation and environmental conservation descending from various European perspectives, there is also a growing appreciation of deeper non-Western perspectives rooted in visions of the Aboriginal Peoples. In Nunavut, the newest northern territory in Canada, complex interactions occur among competing visions of sustainability, development and justice (Doubleday 2003) and we have addressed the interplay of power expressed through constitutions and through art elsewhere (Doubleday *et al.* 2004). Here, I examine the presentation and representation of the Arctic through the lens of Inuit perspective represented in examples of Inuit art, and reflect on possible meanings for deepening our understanding of environment and of culture as a force for sustainability.

G. Humphrys and M. Williams (eds.), Presenting and Representing Environments, 167-180.

Inuit Art and Cape Dorset

The intent here is neither to recapitulate the more than fifty years of development of this Cape Dorset Inuit art enterprise, nor to pass judgement on the art or the artists as to the degree of purity or acculturation represented, but rather to consult selected works from these collections in order to read them as *texts* written in a language we may not understand, but respecting the authenticity of the voices present, and recognising that understandings which are drawn from these works may be both individual and universal. (There have been many useful and thoughtful studies of the development of Inuit art in Canada (Mitchell 1997a,b; Berlo 1999) and in Cape Dorset in particular (Blodgett 1991; Blodgett and Gustavison 1993; Berlo 1999)). In addition, many catalogues of exhibitions and collections of Inuit Art from all regions, including Cape Dorset, are available for reference (for example, Gustavison 1994). In taking this approach, I adopt Dewey's view of art "as providing a sense of the whole of something" (Janesick 2003, 48), and of the role of the artist as restorer of continuity "between the refined and intensified forms of experience that are works of art and the everyday events, doings, and sufferings that are universally recognised to constitute experience" (Janesick 2003, 48–49). In this, I am mindful of earlier work where an entry point from Inuit oral traditions, captured in songs, provided a way of making manifest some intangible culturally based perceptions of the relationship between *Inuk* as person and Arctic as 'home' rather than 'environment' (Doubleday 1999).

The intent of this act of *reading image as text* is not to critique the authenticity or value, aesthetic or otherwise of the works themselves, but rather to look through them as lenses or windows into an arctic world that cannot be assumed to be familiar. In doing so, I expect to catch reflections of the world at my back as well.

Inuit art has achieved recognition on the world stage as a result of its unique aesthetic, its origin and, to some extent, its political utility. Under Inuit leadership, Inuit identity is expressly articulated within the goals for political development in Nunavut, expressed in the Bathurst Mandate. The values guiding the political development of Nunavut have been explored in detail elsewhere (Doubleday 2003) and are summarised here by reference to the four goals ("healthy communities", "simplicity and unity", "self-reliance" and "continuing learning") and related principles enshrined in the Bathurst Mandate titled "Pinasuaqtavut: that which we've set out to do" (Bathurst Mandate 1999).

The Bathurst Mandate reads, in part:

> We believe that: The health of Nunavut depends on the health of each of its physical, social, economic and cultural communities, and the ability of those communities to serve Nunavummiut in the spirit of *Inuuqatigiittiarniq*; the healthy inter-connection of mind, body, spirit and environment … To achieve the dreams of Nunavut we all need to listen closely and learn well in order to acquire the skills we need to increase our independence and prosperity. (Bathurst Mandate 1999)

Table 11.1 *Nunavut and the Goals of the Bathurst Mandate*

Goal	Qualities associated	Strategies identified
Healthy communities	• healthy interconnection: mind–body–spirit–environment	• Inuit *Qaujimajatuqangit* (that is, "Inuit Knowledge")
Simplicity and unity	• government: open, responsive, accountable • public participation	• Inuit Qaujimajatuqangit • fair, understandable, accessible programmes and services • cooperation
Self-reliance	• active participation • build on strengths	• incorporate traditional activities and values into economic development • emphasise unique attributes of people, communities, environment and economy
Continuing learning	• respectful pride • culturally appropriate education	• land and language skills • Inuit Qaujimajatuqangit

The articulation of Inuit identity in art and political life, I argue here, flows from the same, shared spring of relations between Inuit culture and 'land' and is captured within '*Inuuqatigiittiarniq*'. Table 11.1 presents the goals of the Bathurst Mandate, and the qualities and/or strategies identified with them.

In each instance, there are clear linkages between the elements of Inuit culture connected to traditional life on the land and the goals envisioned for the future of Nunavut. The implementation of *Inuit Qaujimajatuqangit* (Inuit knowledge) is a central strategy related to the elements of public life, economic activity, social and cultural practices, environmental opportunities and the health of Nunavut.

Perhaps most significant is the first, where healthy communities are understood to require the "the healthy inter-connection of mind, body, spirit and environment" (Bathurst Mandate 999). I would suggest that, immediately, we are invited to enter a new paradigm for development, and that the Bathurst Mandate, as a statement of core Inuit values, has positioned itself to become the context for sustainability. This point of difference stands in contrast with the liberal democratic values of the moratorium on development recommended for the Western Arctic by the *Berger Report*, which called for a ten-year long delay to be imposed on development. The purpose of the moratorium was not necessarily to redirect development of oil and gas or to question its ends, but rather to allow time for the rights of the peoples of the territory to be addressed formally through negotiations, so that they could be recognised among the beneficiaries of the development.

The recognition of frontier and homeland of Berger is intimately familiar from the perspective of dichotomous processes of analysis. However, not withstanding the utility of the construction of the 'other' as point of contrast, and/or in a dark outcome, eventual domination (Said 1994), there are alternative possibilities waiting to be born. I continue to probe beyond contrast and dichotomy, looking for a way into alternative, and, perhaps, holistic perspectives in northern cultural expression of northern realities, in this chapter, through the entry point of Inuit presentation and representation in art.

Other researchers have also struggled to penetrate the invisible links between art and context, some through engaging the artist as agent (Mitchell 1997a, 14), others by looking to archival sources or even to astronomy (Olson *et al.* 2004) as aids to interpretation. While consultation with the artists themselves may reveal layers of meaning not readily perceived beyond the circle of intimate experience (Mitchell 1997b), it is still possible to anticipate that the art itself has the capacity to communicate both the experience as well as the cultural and environmental context of the artist. Here, I argue that recognising the art *itself* as a gestalt implies that the artist need not articulate fully *in words* the many meanings embedded in the artwork for these multiple meanings to be capable of being transmitted. Rather, I argue that the audience, in our case the viewer, has responsibility for comprehension as well. Mitchell (1997b, 7–8) makes the point that within culture, expressed within language-defined contexts, Inuit artists have shared terminology that is not necessarily accessible to non-Inuit. As an example, she cites Tulugak's views:

> … there are three words commonly used to describe a good piece of sculpture: *inismatsiatuk*, which means 'everything in its perfect place', *takuminuktut*, which means 'that which wants to be looked at', and *suligasuatuk*, which means 'striving for truth'.

Clearly it is a remote possibility that the urban collector of Inuit art in a major American city will share the terminology or the perspectives of the Inuit artists in this example. However, an assessment will inevitably be made across the cultural gap reflecting the capacity of the observer to comprehend. If Dewey's view is correct, what is then comprehended will carry inevitably a grain of the truth of the arctic world that engendered the artist and the art. In this sense, I suggest that the art itself and the experience of the artist and of the viewer constitute a microcosm of the arctic world: that in essence, the Arctic is holographic in relation to Inuit art, and Inuit art constitutes a hologram of the Arctic.

As a strategy for 'grounding' this investigation, I take responsibility for making particular choices from among the multiple, available, representations of Inuit art. Here I elect to examine fewer works more closely, to choose works made by Inuit artists of Cape Dorset, and to reference only those works available for viewing through web-based facilities, and thus accessible to the reader. I accept the liability that attaches to selectivity and make no pretension to wider engagement beyond Inuit art as evidence that carries meaning in support of the argument. This is not intended to make claims for a naïve decontextualisation of art, but rather to illustrate a means of moving from a foundation of the articulation of difference in political and historical meaning, through the contemporary presentation of Nunavut as a multidimensional entity, capable of containing meaning, and of being 'unpacked' ('unpacked' in the sense of "search for the meanings and memories stored inside", given in Phillips and Steiner 1999, 19) from the perspectives of Inuit identity, colonial power relations and intercultural negotiation of meaning (Doubleday 2003); and culminating in a proposal for the possibility of an alternative perspective of the 'land', or 'natural environment' or 'nature', being one entwined with culture. In order to articulate this proposal, I rely on a process of reading together selected aspects of the presentation of values-oriented policy dimensions of Nunavut with representations of aspects of life in the Arctic in Inuit art selected from among the publicly and digitally available works by Inuit artists of Cape Dorset. In this way, the symmetry of politics and art as creative processes embodying values and alternative perspectives can be explored.

The consequences of positioning 'environment' as a category of inquiry, external to the perspective of the actor, and capable of being treated as 'other', are legion: dualism; dichotomy; loss of meaning, and arguably, loss of relationship must all follow.

When we speak of 'the natural environment', do we then inevitably also think of an 'un-natural' environment? While we may use terms such as 'urban' or 'developed' in order to differentiate landscapes dominated by human works from those where such domination is less visible, the viability of such a distinction is increasingly coming into question in some (but not all) geographical writings (Castree 2001; Gregory 2001). In the context of the Arctic, does the positioning of a series of picturesque *inukshuit* on the arctic horizon really differ from the march of pylons in Northern Quebec carrying electricity from generators to cities as far away as New York? Both, after all, are concerned with communication across distance in some respect. However I must quickly add that this observation is intended to be provocative rather than persuasive, and to problematise assumption and interpretation rather than to endorse. In the case of the Arctic, it becomes clearer that categories, whether of nature or of art, are projections, and as such must be examined in a manner that is sensitive to context if meaning is to be apprehended.

Theoretical Perspectives on the Power of Worldview

This is not the place for an exhaustive review of what must necessarily be a complex theoretical formation, however a brief explication of some of the theoretical assumptions driving this inquiry into identity and sustainability is likely to be useful.

The ecological appropriateness of the dominance of Western 'worldviews' and values has been questioned from various critical perspectives, including those that can be described as post-modern critical thought, liberal humanism, post-structuralism, anti-foundationalism and non-Western and international human rights frameworks, to name a few. In turn there has been a renewed interest in reinforcing ideas of the pre-eminence, if not actual dominance, of Western perspectives, as asserted in political and academic discourses. Writers of North American and world environmental history, such as Diamond (1997) for example, have constructed pictures of intellectual evolution using arguments from biology, genetics, paleoecology, and textual materials, which adopt the hierarchical ascendancy of Western agrarian ideals as the logical outcome of a belief in the superior endowment of particular groups, combined with environmental and/or genetic lottery winnings. Others, such as Eisenberg, (See, for example, Eisenberg's (1998, 310) treatment of Kayapo "resource islands" in the rainforest) display a deeper sensitivity to local ecological underpinnings, and as a result can derive meaning in particularities, where a broad brush would only mask the truth of locality. In the context of alternatives to dichotomies, the contrast carries through: Eisenberg (1998, 311–312) understands that nature and culture form continua in specific places, and Diamond's approach perpetuates questions of difference and ascendancy, reinforcing a dialectic of difference and, ultimately, in my view, of conflict.

Previously, I have attempted to resolve issues produced by the constructed nature/human dichotomy by turning to non-Western human societies in search of conceptualisations of relationships that avoided dependency on hierarchy, domination, or social Darwinism. Then, having deciphered the qualitative cultural dimensions *in situ*, I looked for the Western intellectual frameworks capable of accommodating the variability of non-Western experiences of ecology and grasping the fundamental qualitative differences in searching out the threads that weave the web of relatedness. From this analysis, then, the epistemological implications of such a framework in the context of traditional ecological knowledge could be explored (Doubleday 1993).

One of the crucial steps in the intellectual process I have identified, in this shift away from dogmatic approaches dependent on dichotomous views, is adapted from a spectrum of relationship articulated by Jantsch (1975). This step is necessary in order to frame a progression in the recognition of relationship with the non-human ecological, seen as a moving away from instrumentalism embedded in a mythic objective reality to a holistic appreciation of the unity of life, and the consequences of such a vision for policy, action and decision making. Future steps anticipate the application to transcendent understanding of this unity of life as extending not only through time and space, but also as perhaps conveying an innate sense of the sacred.

However, it is one thing to understand nature and culture as a continuum (Dasmann 1976; Weeks 1986), it is quite another to attempt to see nature whole and interconnected, and then to envision how human agency might exercise sustainably from within such a vision. Setting aside the question of the relevance of the great Eastern and Western religious traditions, within geographic writing we find other alternatives addressing this concern, in part through choices of organising principles and strategies of analysis. One example is actor-network theory (Castree and MacMillan 2001), and another is through theoretical bridge-building (following Derridian deconstruction of edifices of "knowledge") in the construction of epistemology/ontology in Western metaphysics (Castree and Wainwright 2001). However, if we are to find a door through which to enter alternative perspectives of the natural environment or a meaningful metaphysics through which to examine a proposal for holographic vision, I argue that we must find a way to begin outside the conventional frameworks of dialectic and hierarchy that imbue contemporary critical theory in geography and related social sciences. This step is a precursor to the questioning of assumptions necessary to attain habits of mind that can consciously initiate sustainable practices, that are rooted ecologically, and that operate multidimensionally within a holographic vision. To avoid confounding assumptions and for ease of examination, it is preferable to begin, if possible, outside and independent of Western traditions of knowledge. This, then, is one potential contribution of evidence of Inuit identity as worldview, expressed in art, and applied to the establishment of alternative perspectives in the presentation and representation of natural environments examined here.

Within contemporary geographical writing, others retain cautious optimism for a continuation of earlier traditions of conservation and wilderness protection within development (Atkinson 2003). These approaches can be understood as parallel in origin to the artistic traditions of representative landscape painting familiar to Europe and to North America. If, however, we seek deeper understanding of the political ecology involved it is necessary to consider both the nature of ecological reality and the hegemony of the fundamental intellectual framework invoked in such analyses, as potentially confounding factors.

Accepting, for the sake of argument, that intellectual frameworks exist as both artefacts of a particular intellectual heritage, (such as the dialectic discourse of Western epistemological tradition), and as a response to experience, (rather than sheer environmental determinism), the challenge then becomes one of obtaining a vantage point from which critical perspective may be obtained. In this way it may become possible to make visible the 'ideological'; thus enabling a new cultural geography that retains traditional concerns for landscape and place while seeking to articulate the ideologies that underpin the presentation and the representation of places and landscapes in text (following the argument of Duncan and Ley 1993).

The benefit of a 'vantage point' is that it avoids the danger of being dialectically forced into counter-point or opposition. In this way, I seek to avoid romancing the 'other' or creating a straw 'opposite', but hope to enable other discourses whose starting points are quite different (Lee 2003). By acknowledging the legitimacy of the lens of Inuit identity and the eye of the Inuit artist seen by reflection in selected Inuit artworks, this chapter seeks to give standing to Inuit knowledge frameworks and to reposition the discourse of 'other', whether it is applied to nature, as in the case of Castree's investigation, or to culture (for example, Braun and Wainwright 2001, 54).

By this means of giving standing to a vision largely outside Western tradition, an alternative perspective and a less familiar way of seeing nature and human as whole can be advanced. The consequences of such a vision, and implicitly, of seeing ourselves within it, have significance beyond relations within the Arctic, reaching to the global, and beyond discourses of development in my view, but this larger investigation is beyond the scope of the present chapter.

The Canadian Arctic Experienced

In the Canadian landscape, particularly in the North, there has been a history of drawing distinctions and framing dichotomies based on perceptions of the 'settled' and the 'wild', or 'wilderness' and the rest (Nash 1967; Atkinson 2003). Inevitably, this framing of distinctions has been coupled with a conflation of meanings of 'wilderness' with resource frontier and storehouse of natural resources, often operating to the general disadvantage of the Aboriginal Peoples of Canada (Brody 1987; Mackenzie and Dalby 2003). Berger, in his *Report of the Royal Commission into a Mackenzie Valley Pipeline,* provided a long over-due correction of this misperception, ensuring a delay of pipeline development to allow northerners time to prepare to deal with such development projects in their own way (Berger 1977).

Beyond the dichotomy of frontier and homeland and the meanings projected against the designations, the Arctic remains whole, and still more than the sum of its parts. The reality is that the Arctic is both frontier and homeland and the struggle for dominance reflects a failure of understanding. It is tempting to misinterpret struggle and to assign outcomes when assumptions flow from dichotomous constructions of the North, and caution is advisable before conclusions are drawn. After examining over 4,000 drawings, Berlo, a well-informed art historian, observed:

> The anthropologist Edmund Carpenter (1973) and others have said of Inuit prints that they are "inauthentic" because they maintain a fiction of arctic life that is a century out of date. The vast majority of Inuit prints depict the 'old ways' of hunting, tracking, sewing hide tents and so forth. *I was surprised to find that most unpublished drawings depict the same thing, despite the fact that modern life in the Arctic is a complex cultural stew.* (Berlo 1999, 191, emphasis added)

However, reading through the images and the conclusion drawn, in the light of the values represented in the Bathurst Mandate, it is not really surprising that the 'old ways' appear so prominently, as the strategy of incorporating the traditional as a way into the new is one that is well recognised by Inuit. What Berlo views here as "stew" could also be seen as a community feast, where people bring what they have and take what they want or need.

What is 'fact' from one perspective is open to other interpretations, including those of the more rarely heard voices of Inuit and Inuvialuit, such as David Rueben Piqtoukun, who expresses another perspective on presentation and representation of the Arctic and of his life within it: "My world cannot be destroyed by modern influences. Although indelibly altered, my traditions and culture will survive" (cited in Mitchell 1997, 14).

Berlo comes close to David Rueben Piqtoukun's view, acknowledging the "act of making art" as an act of construction of identity:

> … in a rapidly changing modern world, the Inuit artist insists on the persistence of her own unique cultural identity. The act of making art inscribes and defines a vision of the self that is crucial to Native identity. It signifies power, authenticity and identity within the Native culture. The artist depicts the reality of late twentieth century life in a culture that, while geographically remote, also participates in a global culture.
> (Berlo 1999, 192)

Yet, in what sense can we understand this idea of participation of Inuit culture in globalisation? While the intellectual heritage of the West has included a presumption that in North America the settlers and their invading cultures, founded on the economics of agriculture and the legalities of property rights, were in some sense superior, or more progressive or more advanced, than the Aboriginal Peoples with their territorial hunting and gathering or shifting cultivation, this interpretation is by no means settled, and this view appears now to be subject to testing. Critiques of agro-industrial society condemn as ecologically destructive the acceleration of rates of consumption, including rates of destruction of productive capacity. These destructive tendencies, as evidenced by system failures such as Bovine Spongiform Encephalopathy (BSE), and in the accumulation of endocrine disrupters in vertebrate reproductive systems (Colborn 1996), are not simply because of a structural ability to consume more, but rather because of an organising ethos that identifies increases in production and consumption with societal goals, as positive outcomes of human development. This is a naïve assumption that connects consumption or more crudely expressed, the 'greed factor', with a moral positivist conceptualisation of the 'good'. In contrast the Inuit hunter-gatherer society functions on a declared ethos of sharing within the cultural group: hoarding food and material wealth is one of the most widely and severely condemned ethical failures that Inuit identify as improper behaviour. (For an account of the historical application and limits of this ethos in one area, the west Hudson Bay region, see Fossett 2001, 206. However, it should be borne in mind that the perception of proximity and the consequent delimiting of the bounds of sharing have been globalised by telecommunications: when the Ethiopian famine occurred in the mid-1980s, Inuit in Canada comprised the largest single Canada aid donor, because, according to the Inuit leadership, they recognised the plight of the starving people from memories of their own experience with starvation in the Keewatin (now the Kivalluk Region) in 1950 (Harrington 1981).) Participation in globalisation can most simply be understood as occurring through transmission between cultures.

Reflecting on the words of David Rueben Piqtoukun again: "Although indelibly altered, my traditions and culture will survive" (cited in Mitchell 1997a, 14): despite the encroachments and negotiated transactions, including the participation in globalisation, Inuit identity persists. Despite the commodification of Inuit art, it retains its capacity to communicate. The question then becomes one of the receptive capacities of the viewer: choices among meanings exist, levels of interpretation are real, and making sense is an active process.

How critical is it then, that precise meanings are carried transparently, by art, between creator and receiver? Indeed, I would argue, even the creation of awareness of limits to understanding, of meanings unclear, of secret knowledge not shared, would serve as an important flag marking the boundaries of assumptions about knowing, and leading to reflection, if we were sufficiently wise.

The World in a Grain of Sand, The Universe in Two Dimensions? Inuit Prints from Cape Dorset

In order to examine the proposition that a work of art, in this case a lithograph (two-dimensional print), could serve to present and represent the Arctic, I propose a reading of images or visual representations as *text,* in conjunction with elements of Nunavut government policy, represented here by the goals of the Bathurst Mandate. In order to explore this approach, I refer to three print works from Cape Dorset, available to be viewed on-line (http://collections.ic.gc.ca/cape_dorset/). The first of the lithographs selected to be viewed/reviewed is by Kenojuak Ashevak of Cape Dorset, titled *Nunavut, May 1992* (Class no.: CD 15PR92 1; Size: 94.5 × 94.5cm; Description: Lithograph). It can be accessed at the Industry Canada/SchoolNet website (http://collections.ic.gc.ca/cape_dorset/artist1/1sli5.html). *Nunavut, May 1992* is a sweeping, exquisite 360° vision, challenging the viewer to determine perspective immediately. It can be understood as a view from below, looking skyward, and seeing sun and moon. It can also be seen as a shaman's view, looking down and through to a reflected sky. It is both poleward-looking and circumpolar, planar and vertical. It allows the viewer to look out of and into the landscape. It contains layers open to interpretation, displaying a body of water that is also both the winter night sky, and the night-less summer day. It is a circular 'Book of Hours' in which time is cyclical, where the multiple seasonal interpretations of the landscape are present visually in a circle of land, ice and water, rimmed with snowy peaks. For three seasons, the land and ice merge, with the open water of summer representing the fourth. The sky/sea holds sun and moon and stars, as well as birds. The lives of the people and of the other living creatures unfold on the sea ice, in the water, or on the land. It is a comprehensive work, richly detailed, and full of textures.

The second illustration is by Jamasie Teevee, and is titled *Summer Travellers* (Class no.: CD222PR72 48; Size: 31.6 × 40.0 cm; Description: Copperplate engraving). It provides many details of the summer travel on land by Inuit, carrying packs and assisted by dogs also carrying packs. The clothing of the Inuit is finely detailed, showing men, women and children in clothing appropriate to their age, gender and status. The dogs appear tiny, perhaps reflecting the reduction in the importance of their role in summer, when, in the absence of snow, sleds could no longer be pulled. The perspective is more conventionally vertical, and the party is clearly in procession, moving away from the viewer. Time is a point at which the figures are frozen, almost as in a photograph, yet the context of travel through the landscape is unmistakable, even though the landscape itself is undifferentiated.

The third illustration is by Pitseolak Ashoona, and is titled *Perils of the Sea Travellers* (CD 24PR60 50, no other information is given). In terms of perspective, this image offers a perspective that is not quite vertical, yet not fully planar. The Inuit in the *umiaq* are surrounded by predatory creatures, perhaps animal, perhaps spirit, with teeth, tusks, and claws outstretched toward them. One creature has even appeared on board the *umiaq*, facing the bowman. The over-arching message is one of ever-present danger, existing in multiple forms, and lying in all directions. The view is one that crosses perspectives, in the sense of seeing from above, but not a clearly birdlike vision, yet not definitely at human height, while also seeing as visible the hunger or desires, whether spiritual or physical, of the forms as well as their physical manifestations as living beings. Here, the properties of perspective and of vision are of a qualitatively different order, bridging the known and the unknown, within a frame of fear. Again, the environmental context, here the sea, is so omnipresent that it is unmistakable even though undifferentiated.

These last two visual 'texts' tell of lives lived, depicting sea travel and summer camping trips, aspects of Inuit life on the land that persist. The details show hunting implements, clothing design, animal domestication, social order, and acute awareness of potential threats to survival. Most striking is the presence through 'white space' of the land and sea. This is reminiscent of an experience I had in 1980, when Inuit leaders functioning internationally initially saw no need to name 'environment' explicitly in their work: it was held to be so obvious that it was explicit without labelling. However, if we compare *Perils of the Sea Tavellers* and *Summer Travellers* with *Nunavut, May 1992*, it creates a curious effect not unlike that of placing photographs against a map of a world. Whereas in *Summer Travellers* and *Perils of the Sea Travellers*, the capacity to define rests more heavily with the viewer, the richness and complexity of *Nunavut, May 1992* combined with its remarkable multiple perspectives serves to challenge any inclination an observer might have to divide this work: it demands to be seen first as a whole before any subdividing can be performed. This observation opens the door to attaining the objective that was the primary point of departure for this investigation: the pursuit of a way of seeing that is capable of comprehending the natural environment whole, including humans within the frame. Surely this is in the spirit of *Inuuqatigiittiarniq,* the healthy inter-connection of mind, body, spirit and environment, that is central to the Bathurst Mandate, and to Inuit aspirations for Nunavut?

From this small beginning, it appears possible to find alternatives – perspectives able to both envision the natural landscape and the humans within it – and to make visible a vision of the whole. In other words, it seems that Inuit art can be read as presenting and representing an interconnected, multidimensional, and inevitably ecological, arctic 'text'.

What then are the implications of establishing a possibility of *holistic vision* as *perspective*? What does this mean? As Williams and Humphrys (2003, 2–3) astutely point out in the context of citizenship education, power over territory is likely to determine content in the absence of participant control over context. Similarly, the alternative meanings of the arctic landscape will inevitably be given content by the contexts of the observers, and in the absence of active control by Inuit, the exercise of power will be by those situated in the boardrooms of empires contingent on the perpetuation of the peripheries, whether as frontier, or wilderness or resource storehouse. So Nunavut, whether seen as a public government, as a bundle of private rights determined under a comprehensive land claim, or as a vast land of great beauty and great wealth, bears a significant burden. Can Nunavut, as presented in the principles of the Bathurst Mandate, become the defining context in the face of competition with the forces of economic globalisation which seek to reduce it to categories of utility? Or in the alternative, if the struggle for the right of Inuit to give content to the context of Nunavut cannot prevail, (again following Williams and Humphrys 2003), is it not then possible to conceive of Nunavut as 'person', much as the multinational corporations that seek its wealth are?

Can this moment of insight, however brief, when perhaps "the doors of perception are cleansed and all things can be seen as they are – infinite", (to follow Huxley 1954) be seized? Or will this presentation by Inuit, through representation in Inuit art of an alternative vision of a way of being in the world, quickly be dismissed – as novelty or curiosity, or as anachronism – belonging to another era? Any relevance to sustainability, or to peace, or to a dream for the future in Nunavut that does not recapitulate the past, is likely to be ignored or actively disputed. It may even cause acceleration of assimilationist strategies designed to sever the organic threads embroidering the tundra tapestry and sewing the world of sea ice to migratory seabirds, to humans as hunters and mothers, and to future generations, by imposing the calculus and the categories of resources and jobs, and that first feed the boardroom tables. (Why is it that if the free do not become slaves, they so often must become miners? Here I recall the almost prescient views of the former Minister of Economic Development, Gordon Wray, who saw mining as a driver of the northern economy and a source of employment for Inuit.)

But beyond pessimism there is a glimmer on the far flat horizon of the Arctic. If, as Thomas proposes, "the well-being of the planet as a whole, becomes the new *StaatsIdee* for a future global citizenship" (in Williams and Humphrys 2003, 264), and Canada is to take leadership, then surely we are justified in attempting to think differently and to see differently? Then, perhaps, alternative perspectives, illuminated by Inuit art, and perhaps by Inuit values, can be seen as potential sources of metaphor for interdependency, for interconnectedness, for seeing a living planet and for learning to tend it well?

Canada is thought by many to be uniquely positioned. According to Kofi Annan, Secretary General of the United Nations, Canada can be seen as:

> a kind of United Nations unto itself because it is such a diverse and multilingual society of aboriginals, francophones, anglophones and immigrants … a born multilateralist, naturally adept at the give and take of international cooperation. … It is your ability as Canadians to look at the needs of others, to feel compassion for their suffering as part of your own, which speaks to the best of you as Canadians.
>
> (Kofi Annan 2004)

This too, is a form of holistic vision.

In the Arctic and elsewhere, we have had the gold rush, the carpetbaggers, the parks movement, the conservation movement, the wilderness movement, and now the sustainable development movement – but all movements establish identity through groups and interests, further splintering a reality that must be seen whole to be fully understood. Inuit have articulated and presented, in representations in art and in policy, a different vision. Perhaps this alternative will assist more of us to see differently, to understand ourselves as responsible, and as therefore entitled, if not obligated, to engage. If we see ourselves as becoming self-governing, and understand this process as iterative, on-going, continuing learning toward a possibility of seeing the whole beyond the parts, "in the spirit of *Inuuqatigiittiarniq*; the healthy interconnection of mind, body, spirit and environment", then it is not only the Inuit of Nunavut who are accountable:

> To achieve the dreams of Nunavut we all need to listen closely and learn well in order to acquire the skills we need to increase our independence and prosperity. (Bathurst Mandate 1999)

References

Annan, Kofi. (2004). Interview. *The Ottawa Citizen*, 9 March.

Atkinson, K. (2003). Wilderness and the Canadian mind, *British Journal of Canadian Studies*, 16, 228–242.

Bathurst Mandate (1999). The Bathurst Mandate, 1999. Government of Nunavut. www.gov.nu.ca/Nunavut/English/departments/bathurst/ (accessed November 2004).

Berger, T. R. (1977). *Northern Frontier, Northern Homeland*. Vol. 1, Ottawa: Minister of Supply and Services.

Berlo, J. C. (1999). Drawing upon the past: negotiating identities in Inuit graphic arts production. In R. B. Phillips and C. B. Steiner (Eds.) *Unpacking Culture: Art and commodity in colonial and post-colonial worlds*. Berkeley: University of California Press.

Blodgett, J. and Gustavison S. (1993). *Strange Scenes: Early Cape Dorset drawings*. Kleinburg, Ontario: McMichel Art Collection. Cape Dorset Revisited, Exhibition Catalogue.

Braun, B. and Wainwright, J. (2001). Nature, postructuralism, and politics. In N. Castree and B. Braun (Eds.) *Social Nature: Theory, practice, and politics*. Oxford: Blackwell.

Brody, H. (1987). *Living Arctic*. Vancouver: Douglas and McIntyre.

Castree, N. (2001). Socializing the natural. In N. Castree and B. Braun (Eds.) *Social Nature: Theory, practice, and politics*. Oxford: Blackwell.

Castree, N. and MacMillan, T. (2001). Dissolving dualisms: Actor-networks and the reimagination of nature. In N. Castree and B. Braun (Eds.) *Social Nature: Theory, practice, and politics*. Oxford: Blackwell.

Colborn, T. with Dumanoski, D. and Myers, J. P. (1996). *Our Stolen Future: Are we threatening our fertility, intelligence, and survival? A scientific detective story*. New York: Dutton.

Dasmann, R. F. (1976). Ecosystem people, *Parks*, 1, 5–7.

Diamond, J. M. (1997). *Guns, Germs, and Steel: The fates of human societies*. New York: W. W. Norton and Co.

Doubleday, N. C. (1993). Finding common ground: Natural law and collective wisdom. In J. T. Inglis (Ed.) *Traditional Ecological Knowledge: Cases and concepts*. International Program on Traditional Ecological Knowledge, International Development Research Centre. Ottawa, Canada.

Doubleday, N. C. (1999). Arctic worlds and the geography of imagination. In A. Buttimer and L. Wallin (Eds.) *Nature, Culture, Identity*. Dordrecht: Kluwer Academic Publishers.

Doubleday, N., Mackenzie, A. F. D. and Dalby, S. (2004). Reimagining sustainable cultures: Constitutions, property and art, *Canadian Geographer*, 48, 389–402.

Doubleday, N. C. (2003). The nexus of identity, Inuit autonomy and arctic sustainability: Learning from

Nunavut, community and culture, *British Journal of Canadian Studies,* 16, 297–308.
Duncan, J. and Ley, D. (Eds.) (1993). *Race/Culture/Representation*. London and New York: Routledge.
Eisenberg, E (1998). *The Ecology of Eden*. New York: Alfred A. Knopf.
Fossett, R. (2001). *In Order to Live Untroubled*. Winnipeg: The University of Manitoba Press.
Gregory, D. (2001). (Post)Colonialism and the production of nature. In N. Castree and B. Braun (Eds.) *Social Nature: Theory, practice, and politics*. Oxford: Blackwell.
Gustavison, S. (1994). *Arctic Expressions*. Kleinburg, Ontario: McMichel Art Collection. Exhibition Catalogue.
Harrington, R. (1981). *The Inuit: Life as it was*. Edmonton: Hurtig Publishers.
Huxley, A. (1954). *The Doors of Perception*. New York: Harper.
Janesick, V. J. (2003). The choreography of qualitative research design: Minuets, improvisation and crystallization. In N. K. Denzin and Y. S. Lincoln (Eds.) *Strategies of Qualitative Inquiry*. Thousand Oaks, London: Sage Publications.
Jantsch, E. (1975). *Design for Evolution: Self-organization and planning in the life of human systems*. New York: G. Braziller.
Lee, C. (2003). *Cowboys and Dragons: Shattering cultural myths to advance Chinese-American business*. Dearborn: Dearborn Trade Publishing, Kaplan.
Mackenzie, A. F .D. and Dalby, S. (2003). Moving mountains: Community and resistance in the Isle of Harris, Scotland, and Cape Breton, Canada, *Antipode,* 35, 309–333.
Mitchell, M. (1997a). Inuit Art is Inuit Art: Part One, *Inuit Art Quarterly*, 12, 4–15.
Mitchell, M. (1997b). Inuit Art is Inuit Art: Part Two, *Inuit Art Quarterly*, 12, 4–16.
Nash, R. (1967). *Wilderness and the American Mind*. New Haven: Yale University Press.
Phillips, R. and Steiner, C. (Eds.) (1999). *Unpacking Culture: Art and commodity in colonial and postcolonial worlds*. Berkeley: University of California Press.
Olson, D. W., Doescher, R. L., and Olson, M. S. (2004). The Moon and the marathon, *Sky and Telescope*, September, 34–41.
Said, E. W. (1994). *Culture and Imperialism*. New York: Random House.
Thomas, P. F. (2003). Dissonance and assonance in the cultural mosaic: Reflections on a Canadian educational problematic. In M. Williams and G. Humphrys (Eds.) *Citizenship Education and Lifelong Learning: power and place*. New York: Nova Science Publishers, Inc.
Weeks, N. (1986). National parks and native peoples: A study of the experience of selected other jurisdictions with a view to co-operation in northern Canada. In H. Beach (Ed.) *Contributions to Circumpolar Studies*. Uppsala Research Reports in Cultural Anthropology, 7, 83–152.
Williams, M. and Humphrys, G. (Eds.) (2003). *Citizenship Education and Lifelong Learning; power and place*. New York: Nova Science Publishers, Inc.

12. CULTIVATING A NEW CATTLE CULTURE: LIFELONG LEARNING AND PASTURE LAND MANAGEMENT

IAN MACLACHLAN, NANCY G. BATEMAN AND
THOMAS R. R. JOHNSTON

"I suppose it would be better for the wildlife if we weren't here at all. But we are here and I think there's a place for beef cattle in this environment".
Bob Dobson (*Canadian Cattlemen* 1995, 15)

'Pasture land management' implies the planned and sustainable use of a grass-covered landscape by humans to feed and accommodate livestock. Pasture land may be found in a true 'grassland ecozone' that characterises much of North America's Great Plains or it may be tame pasture that has been carved out of Canada's boreal forest. In the latter case, grass is not the natural climax vegetation but it grows well where the natural forest cover has been cleared by human hands. Thus pasture land (a term used interchangeably with 'grazing land') is a social construction, created both as a conceptual land use category and as an area of land that supports grazing activity. Grazing implies feeding domestic livestock such as cattle and sheep on standing grass although there is always some competition for the grass resource from 'wild' animals. Pasture land is an economic land use, driven by a profit motive to nourish animals and create value in human terms. Given the artifice and self-interest that is an integral part of the pasture land concept, it may seem paradoxical that sustainable pasture land management is concerned with restoring the biotic diversity and biomass productivity in a tangible and deliberate mimesis of the natural climax vegetation that once characterised the grassland and boreal forest ecozones. But we must also acknowledge that the notion of a pristine pre-industrial landscape is no less a social construction than large-scale factory farms, or for that matter, current models of sustainable agriculture. Pasture land is a constructed landscape that has been subjected to human interventions that range from aggressive clear-cutting and a total change in vegetative cover to the extensive use of rangeland in which it is animal activity, mediated through stocking density, that has had the greatest impact on the ecosystem.

This chapter illustrates how views about pasture land management have developed in Canada. Farmers and ranchers are growing aware of the environmental

G. Humphrys and M. Williams (eds.), Presenting and Representing Environments, 181-195.

hazards posed by livestock grazing and they are looking at the landscapes they manage from an increasingly ecological perspective. They are applying new techniques to increase biodiversity and to reduce the impact of ruminant grazing on both upland and riparian zones. This process, which includes both the acquisition of knowledge about grassland ecosystems and changes in methods of grazing management, is interpreted as a new cattle culture driven by a process of lifelong learning, leading to positive environmental outcomes that imitate the diversity and quality of the natural environment.

The chapter begins with a discussion of the importance of cattle grazing as a contemporary land use and as a potential agent of environmental degradation. The second section of the chapter reviews the literature describing the effect of various programmes intended to encourage sustainable agricultural practices. The third section deals with lifelong learning applied to agriculture, as well as a selective review of efforts to transmit knowledge and induce behavioural change in those making grazing management decisions. To illustrate these processes, two case studies are presented. The first case study features informal learning by an Ottawa Valley farmer who is leading by individual example and sharing his methods with others. The second case study concerns a more structured but voluntary programme called 'Cows and Fish', operated by the Alberta Riparian Habitat Management Society, which works with community members across Canada to increase awareness of riparian zone management.

Cattle Grazing as Land Use and Agent of Environmental Change

Agricultural activity is an extensive land use, subjecting a larger area of Canada's ecumene to the environmental impact of human activity than any other economic sector. Cattle production is a vitally important component of Canadian agriculture. Of Canada's 230,000 farms, 30 per cent are classified as beef cattle operations and a further 8 per cent are dairy farms. Canada's 69,000 cattle farms include approximately 500 large-scale feedlots, intensive grain feeding operations that feed grain and silage to cattle confined in pens. Many of these large-scale feedlots have no pasture land thus no grazing activity. Notwithstanding the use of intensive and confined feeding techniques at some times of the year and for some classes of cattle, 38 per cent of Canada's farms rely on the grazing of standing grass as a feed source. Of all agricultural land uses, tame and natural pasture represent 30 per cent of Canada's agricultural land area, ranging from a low of about 10 per cent in Prince Edward Island and Quebec to a high of 56 per cent in British Columbia (Statistics Canada 2004). With this considerable land area subject to grazing activity, pasture land management plays a key role in the environmental sustainability of Canadian agriculture and more broadly in Canada's ecumene as a whole. The land area subject to grazing is vast. Cattle farms are dispersed across the country from the Avalon Peninsula of Newfoundland to Vancouver Island. Just as grazing affects a very large land area, any effort to promote sustainable practices or to enhance lifelong learning about pasture land management must be conveyed over a very large area. This presents an administrative and communications challenge to information dissemination and technology transfer.

Contemporary concerns over the environmental impact of cattle grazing have a long pedigree in Canada. The British landscape painters of the Romantic period offered bucolic images of rural landscapes featuring pastoral scenes of livestock grazing peacefully on rough pasture, providing a counterpoint to the environmental horrors of an urban-industrial age. In Canada, the link between domestication, animal husbandry and the rustic charm of rural landscapes was decisively broken in the late nineteenth century when the folly of Texas-style cattle ranching was first exposed on the Prairies. Alberta's early ranching industry in southern Alberta was an extension of a cattle culture and stock-raising technology that may be traced to the Gulf coast of Texas (Morton 1938, 91). And, as in Texas, overgrazing mixed with periodic climatic calamities such as drought and winter kill signalled that an extensive and unregulated ranching system was unsustainable. In 1899, the Commissioner of Public Works for the North-West Territories observed: "in the Pincher Creek District, which was at one time looked upon as one of the best grazing areas in the Territories, the range has become so eaten out owing to want of any regulations regarding the grazing of cattle at large, that some of the Ranchers there have had to move their cattle away to other and less crowded portions of Alberta" (Ross quoted in Breen 1983, 120). Jordan (1993) has argued that while climatic conditions alone would have eventually destroyed the Texas ranching system, overgrazing and competition with crop agriculture and sheep ranching in the U.S. hastened the demise of ranching on the open range. Free grass encouraged over-stocking in the rangeland states and so damaged the short-grass prairie that there was nothing left for cattle to eat. Jordan (1993, 239) concludes: "The Texas system was not merely maladapted to the Great Plains; it was not sustainable in any environment and would have collapsed even in the lushest and mildest of settings".

Contemporary grazing management has responded to the environmental fiascos of the late nineteenth and early twentieth centuries. Stocking densities are more closely controlled and monitored by grazing leases, stock raisers can reach their animals and deliver winter feed when it is required, and water can be supplied more reliably than was the case on the open range. From the crown owned rangelands of Alberta and British Columbia to the privately owned tame pastures on the mixed farms of eastern Canada, cattle grazing is perceived as a comparatively benign agricultural land use, unlike wheat monoculture or intensive livestock operations. Perhaps because pastoralism using standing grass seems like such a natural and traditional use of land, there has been a tendency to underplay its environmental impact until comparatively recently.

However, the environmental consequences of livestock grazing have attracted substantial scientific interest in recent years (see, for example, Fleischner 1994), and have led to mounting conflict within rural communities (Chilson 1997). The environmental threat from cattle grazing is becoming a contentious item on the policy agenda of rangeland states and provinces. Evans has described the 'new range wars' in which the overgrazed public lands of the western United States are described as 'cow burnt.' Environmental activists argue that, "Livestock ranching is the single most ecologically damaging activity we engage in" and that its environmental impact should be equated with clear-cut logging and strip mining of coal (Evans 2001, 69). The growing concern about the environmental impact of livestock has not been confined to the west. Increasing accumulations of soil nutrients and the threat they pose to water quality in Quebec and Ontario point to a more generalised need for improved livestock management practices (Commissioner of the Environment and Sustainable Development 2001, 145).

Yet, by incorporating fundamental grazing principles – seasonality, intensity, timing, rotation and stocking rates – modern-day cattle grazing management can fill the ecological niche vacated through the extirpation of the plains bison during the nineteenth century. Bison are recognised as one of the key influences in the development of North America's grassland ecozone (Bradley and Wallis 1996). Unlike bison, that wandered over an extensive territory of thousands of square kilometres on an annual basis, domestic cattle are now more closely managed and their movement constrained by fences. Thus, cattle grazing may be managed as an act of environmental mimesis, in which a domestic species is managed to imitate the impact of an indigenous species. The promotion of such mimetic practices points to the need for an improved understanding of the link between lifelong learning about the environment and motivating improved grazing management practices among cattle producers.

The Effect of Financial Incentives and Promotional Programmes for Agriculture on Environmental Quality

Considerable research has evaluated the factors associated with the adoption of sustainable and environmentally responsive agricultural practices. Brotherton (1989; 1991) classifies participation in agro-environmental programmes under "farmer factors" (characteristics of producers that predispose them to participate) and "scheme factors" (attributes of the programme that encourage participation). As an example of the former, farmers using environmentally sustainable practices in the United States tended to be younger and had higher incomes than conventional farmers (Comer, Ekanem, Singh and Tegagne 1999).

Wilson (1997) used Brotherton's (1989) classification to assess participation in the Cambrian Mountains Environmentally Sensitive Areas (ESA) scheme in central Wales. The ESA is considered to be the "flagship of agri-environmental policy" in the European Union, yet the principal motive for participation appears to have nothing to do with the inherent advantages of sustainable practices or any ethical commitment to the environment. Instead, cash payments were the principal motive for sustainable practices; Wilson's survey revealed that for 69 per cent of scheme participants, financial factors were the primary reasons for joining the programme. Conservation, as an intrinsic motive, was mentioned by only 6 per cent of respondents (Wilson 1997, 77).

While attempts to reverse the environmental damage with state subsidies to promote conservation behaviour have been effective, these efforts tend to produce short-term behavioural change. When the subsidies end, so too do the ameliorative farm practices (Jacobson, Sieving, Jones and van Doorn 2003, 603). Environmental education leading to progressive attitudes and changes in farmer behaviour and agricultural management practices would seem to be more effective in long-run environmental improvement. One step in this direction is to provide informative and pragmatic educational materials to teach farmers how they may make these improvements (Jacobson, Sieving, Jones and van Doorn 2003, 603).

In a New Zealand context, Rhodes, Leland and Niven (2002) surveyed 718 pastoral farmers in Otago and Southland to explore the effect of the provision of information about riparian management. Reassuringly, they found that the level of exposure to information was positively associated with more positive attitudes, greater levels of knowledge, intentions to take actions to improve riparian health and the actual adoption of best management practices for riparian zones. However, all associations were relatively weak and in some cases respondents who received no information had actually adopted quite progressive attitudes and practices. Financial barriers inhibited investment in environmental impact mitigation such as riparian fencing. Thus, the provision of even minor financial incentives could increase rates of adoption of best management practices.

While some level of compensation may be an effective means of priming the pump and getting farmers to experiment with progressive management strategies, it appears that long-term changes in attitudes, behaviour and technique will be required so that sustainable practices remain consistent and integral aspects of management. One approach to inculcating such an outlook could be based on a lifelong learning ethic in which farmers internalise sustainable management techniques and apply them of their own volition.

Cattle Producers, Lifelong Learning and the Information Economy

Although some aspects of the cattle producer's lifestyle seem traditional and proudly based on a family farming legacy that dates to the nineteenth century, cattle raising, livestock husbandry and pasture land management are integral parts of the new economy and are profoundly influenced by the information revolution that continues into the twenty-first century. Cattle producers are deluged with information about cattle prices and market trends, farm financing, pharmaceuticals and neutraceuticals, environmental impacts and of course, grazing land management practices. Based on the enthusiasm and rapidity with which many of these innovations are adopted, agriculturalists are learning about new methods and applying this knowledge in their day-to-day operations.

A growing proportion of farmers have had some form of traditional post-secondary agricultural education which is likely to influence their understanding, acceptance, and rate of adoption of sustainable practices (Comer, Ekanem, Singh and Tegagne 1999).

Table 12.1 *Traditional and lifelong learning models*

Traditional Learning	Lifelong Learning
Teacher as fount of all wisdom	Educators are guides to sources of knowledge
Knowledge reception	Learn by doing
Learn as individuals	Learn in groups
Tests as prerequisites	Assessment to guide future learning
Good students may continue learning, bad students may not!	Lifetime access to learning opportunities

Source: Adapted from *Lifelong Learning in the Global Knowledge Economy: Challenges for Developing Countries,* www.TechKnowLogia.org (accessed October 2004).

However, much of the continuing process of technological change in agriculture may be understood as a lifelong learning process. The left-hand column of Table 12.1 portrays a top-down model of learning in which the agenda is set by the teacher while the student is a passive recipient and performer, progressing incrementally according to a hierarchically imposed system of performance, assessment and promotion. The lifelong learning paradigm represented in the right-hand column approximates the on-the-job training and ethic of continuous quality improvement that characterises the application of knowledge to contemporary agricultural production. Educators in a lifelong learning scenario seldom have the authority to judge and grade learning outcomes while the pedagogical environment is more likely to include informal group learning and a strong applied component in a trial and error process. Trial and error suggests that it is the student who does the assessment and grading, promoting the concept, theories and techniques that appear to work while failing those which prove to be less effective. To some degree, every farm functions as a laboratory and every farmer acts as scientist, student, and teacher.

Expanding on the contrast between traditional and lifelong learning, Table 12.2 presents a two-dimensional classification grid of learning modes. Structured learning modes include formal schooling and the informal, but not less structured learning, that takes place on the farm as an intergenerational transfer of technology, a socialisation process that features learning-by-doing with supervision by parental or community mentors. Unstructured learning may take place under the formal auspices of agricultural extension programmes operated by provincial departments of agricul-

Table 12.2 *Learning Modes Grid*

	Structured	Unstructured
Formal learning	School Vocational training Post-secondary	Extension services Promotional programmes
Informal learning	Socialisation Intergenerational transfers	Mentorship Self-taught Trial and error

ture or by a host of promotional farm programmes such as the Canadian Cattlemen's 'Quality Starts Here' – a programme designed to educate producers and to promote the implementation of on-farm food safety initiatives such as Hazard Analysis Critical Control Point. The programmes are conspicuous by design, with a high media profile. But a less conspicuous form of self-paced and self-driven learning takes place informally in an unstructured environment every time a farmer observes some form of grazing behaviour or pattern of erosion, tries some new technique, no matter how trivial the change or how slight the advance, and then tells someone else about it in a chance meeting at the implement store or over coffee.

It is difficult to assess which of these lifelong learning modes, structured or unstructured, is the most effective in transmitting innovative agricultural practices or to measure which type of information and technology transfer is dominant. As an exploratory step in this direction, this chapter offers two case studies of environmental stewardship to promote sustainable grazing in Canada. The Dobson farm illustrates unstructured and informal lifelong learning by a single farm operator. Cows and Fish, on the other hand, is a structured programme with funding from government departments, conservation agencies and producer groups that is intended to foster lifelong learning and behaviour change with respect to grazing management decisions of thousands of producers throughout Canada.

The Dobson Farm: An Ottawa Valley Case Study in Unstructured and Informal Lifelong Learning and Environmental Stewardship

Bob Dobson is the fourth generation operator of a family farm that has existed at the same location since 1857. Located between Cobden and Pembroke, in the heart of the Ottawa Valley, Dobson's operation is typical of Ontario's mid-sized cattle farms. He owns 250 acres of land, mainly in pasture, and he usually rents about the same amount of land to harvest hay and provide winter feed to his stock. Like many cattle producers, Dobson practises a form of adaptive management, shifting from one product and market segment to another based on his assessment of market trends and best use of the resource base. In 1989-90, Dobson transformed his farm-based business from cow-calf production to building up a calf weaning centre and backgrounding of stocker cattle, feeding them tame grass to build frame size in preparation for high protein grain finishing on an Ontario feedlot. Since that time he has shifted back into cow-calf production and is now developing his business as a specialty grass-fed beef producer. He now feeds his own stock to slaughter weight without the use of supplemental growth hormones or antibiotics and after having the animals custom slaughtered he sells beef direct to consumers in the Cobden-Pembroke area and from a stall at the Carp Farmer's Market in Ottawa's rural-urban fringe.

However, Dobson has become best known for his work in sustainable grazing land management and stewardship of the riparian zone of the creek that runs through his property. He was the first recipient of the Ontario Cattlemen's Environmental

Stewardship Award in 1994. In 2003, he was inducted into Renfrew County's Agricultural Hall of Fame. His farm has been featured in *Harrowsmith* magazine (Webster 1994) and his achievements warranted cover story treatment in *Canadian Cattlemen* (1995), the trade magazine for cattle producers in Canada.

When he took over the farm from his father in 1967, best management practice called for maximum field size to accommodate large farm machines. Many fences, together with the vegetation and animal habitat that develops along fencelines, had been removed, and few trees or shrubs remained. The farm had been made as featureless and denuded of natural vegetation as possible to capitalise on the large-scale monoculture philosophy of the 1960s. Concerned with a shortage of water in the farm's stream during the summer months, Dobson began a long-term programme to regenerate the farm's natural habitat with a variety of trees and shrubs, particularly in the riparian zone. He re-established fencing to subdivide pastures into smaller units required to implement rotational grazing, restoring the smaller fields and linear fenceline strips of vegetation that characterised traditional Ottawa Valley agriculture in the nineteenth century.

Rotational grazing requires a source of water which is accessible to each separate pasture unit and which can provide for the cattle's water needs throughout the year. However, cattle trample bank vegetation, accelerate erosion and, through their manure, contribute a heavy nutrient load to natural watercourses. The nutrients from manure encourage rapid growth of aquatic vegetation that then causes eutrophication and an increase in biological oxygen demand, reducing the capacity of the stream to support desirable species of fish, amphibians and aquatic invertebrates. As part of his rotational grazing plan, Dobson implemented the technique of off-stream watering to provide an alternate upland water supply. He built a thousand gallon (4,500 litre) watering trough of reinforced concrete, situated well away from the stream yet located so that it could water cattle in different pastures. To pump stream water 300 metres to the trough site, Dobson dug a shallow well and installed a floating pump powered by marine batteries that are recharged by a solar panel. The solar-powered distribution system was installed below the frost line and has operated at temperatures of –20°C and even colder with only one change in batteries between 1996 and 2004. The stream is now in much better health than previously, supporting a vigorous community of fish and other aquatic life.

By 2004, Dobson had planted some 20,000 trees and shrubs, selecting species that provide the best and most varied habitat for birds and wildlife. The trees and shrubs also create shelter-belts to block the wind, protect wintering areas for cattle, catch the snow and retain spring runoff. Shrubs such as elderberry were planted between the trees along stream banks and fencelines to provide the variety of habitat required for a balanced ecosystem to develop. Bird species such as loggerhead shrike now control the field mice, impaling their prey on the long spikes of hawthorn trees that shelter cattle in areas that were once open ground. Before and after photographs taken by Dobson as a simple but reliable on-site monitoring tool show the once denuded landscape in the 1970s and how it had been transformed by 2004 (Dobson Farm 2004).

How did Bob Dobson acquire his biodiversity ethic and learn the methods he applied in the award-winning redevelopment of his farm? He is a modest man who does not dwell on his achievements or how he learned the methods that have proven to be so successful. He "read a few things here and there" and learned by doing. This self-directed informal and unstructured learning programme is well illustrated by a comment he made in 1996.

> We used to wean our calves in the barn yard ... in the weaning pen ... We used to treat a fair number of cattle [for illness] like too many, too high a percentage. When – I think somebody suggested to me you know – and maybe it was that year also that we had some real good grass, real good pasture in October. So, it was too late to cut and try to make hay and the only animals I had to use it, ... were these new calves that were coming in and I really was hesitant about letting calves out on to any pasture because we had to treat so many it was too hard to get them in. Once you put them out into a fifty acre field it's too hard to separate one to bring it in, [for treatment]. So, I had a bit of a ... what do they call it, a paradigm shift and all of sudden I thought I would try it. So as soon as we started weaning the calves in the field rather than into the barn yard, ... we immediately reduced the number of animals that got the sniffles and that we had to treat for shipping fever. We had healthier animals and any we treated bounced back the next day, you wouldn't know they were sick. So, once we got them out into the field they were much healthier. So, we now ... put a bit more fertilizer on towards the end of July or the first week of August to get a good flush of growth through the end of August ... And we'll aim for about a hundred and fifty acres of good third or fourth grazed pasture fields for the months of October and November. And they [weaned calves] stay there until the snow comes. I mean we even leave them out there, there's the odd night you feel a little bit sorry for them and I've tried bringing them in, put them in the barn, but they are much healthier out there, even on a night when there are a few ice pellets coming and a little bit of freezing rain coming in November.
>
> (Interview with Bob Dobson, 31 July 1996)

This extended quotation gives a realistic sense of how a farmer develops an idea out of a suggestion with no certainty of whether it will work better than current methods. Dobson developed a new technique for feeding weaned calves on open pasture instead of in the barnyard where they were handy for diagnosis and treatment of disease. The field was a healthier environment so that less observation and treatment was needed. A mixture of a suggestion, a paradigm shift, and a successful experiment led to the development of a new grazing regimen. This is how Bob Dobson learned about pasturing newly-weaned stocker cattle and by using that same trial and error, learn-by-doing process he has developed the many management innovations that contribute to the long-term economic and environmental sustainability of his farm.

He is a firm believer in sharing what he has learned and he now hosts a number of farm tours each year that welcome farmers and other visitors who want to learn about sustainable farm practices.

How Does a Structured Education Programme Foster Lifelong Learning and Behavioural Change in the Ranch Country of Southern Alberta?

Riparian zones are vulnerable to disturbance from cattle (Belsky, Matzke, and Uselman 1999). If access to waterbodies is not restricted, livestock will spend a disproportionate amount of time there, attracted by easy access to water, shade and shelter (Fitch and Adams 1998). This may result in the degradation of the riparian zone's physical form and hydrologic functions and can lead to changes in the composition of both vegetation and wildlife communities.

In Alberta, where the riparian transition zone between upland areas and aquatic systems accounts for approximately 1.3 million hectares of rangeland, a partnership of government agencies, producer groups and conservation organisations combined forces in 1992 to establish what is now the Alberta Riparian Habitat Management Society. Also known as 'Cows and Fish', the programme's goal is to work with communities on riparian awareness, in order to improve the way that riparian zones are managed and to enhance environmental quality and overall ecological function of riparian zones (Cows and Fish 2004a). Cows and Fish is an example of a voluntary resource conservation programme designed to achieve its aims through a combination of education and information provision, along with community and social-capital development initiatives. The 'Cows and Fish Process' comprises five elements – awareness and education, team building, tool building, community-based action and monitoring (Cows and Fish 2004b). Consistent with the lifelong learning model outlined in Table 12.1, the programme employs a variety of tools intended to increase awareness and promote sustainable management decisions in relation to riparian zones. The tools include community presentations given by Cows and Fish staff, riparian demonstration and profile sites, the use of profile producers and other local landowners as leaders of change, riparian health inventories and assessments, intensive field courses, workshops, a series of topic-specific fact sheets, technical liaison to assist landowners, guidance on forming and networking of local community groups, and a core document entitled, *Caring for the Green Zone: Riparian Areas and Grazing Management* (Fitch, Adams and O'Shaughnessy 2003).

It is not yet well understood within the general population that a degraded riparian zone can lead to declining water quality, increased erosion, more extreme flood events and loss of both aquatic and terrestrial habitat (see, for example, Fitch and Adams 1998). Therefore, the first element in the Cows and Fish Process, awareness and education, is meant to enhance ecological literacy about the benefits of healthy riparian systems. Ecological literacy means increasing an individual's knowledge about the ecological functions of riparian zones and supporting it with information on management options that may be suitable for a given site, enabling the individual to develop more positive attitudes and make independently informed decisions about management of that riparian zone. As observed by Bateman (2001, 77-80), ecological literacy is a precursor to the adoption of sustainable management practices and, equally important, continuing to apply informed management decisions over the long-term. This first element of the Cows and Fish Process is based on the premise that acquiring information and positive attitudes are necessary conditions before a change in behaviour can occur.

Team building, tool building and community-based action continue to enhance ecological literacy and are designed to construct local social capital, enhancing the probability that environmentally sustainable grazing and riparian zone management practices will be adopted. Team building by Cows and Fish staff creates partnerships among community members including producers, interested individuals in the community, conservation and community groups, resource managers and local-area authorities. This is achieved by bringing together those who share common interests and perspectives on riparian health in the hope that local networks will emerge. This element is consistent with the view that educators are guides to a learning process as opposed to founts of wisdom (Table 12.1).

Tool building places practical information about sustainable riparian zone management and various grazing systems before producers and resource managers. Here again, programme staff act as conduits, demonstrating on-the-ground examples of management options. These are often hosted by a producer who has implemented a particular strategy, can explain how they decided to incorporate it into their management regimen, and share the benefits and any pitfalls so that others may learn from experience. Fundamental to tool building is an understanding that any strategy can be modified to suit conditions on any given operation.

The illustration and discussion of management options at a demonstration site tour highlights the important link between the group learning scenario concrete applications. According to one producer:

> It becomes a learning thing over time … my own approach to grazing management has changed completely. Seeing what did work, [I was] able to incorporate it into my own operation, to better my place … I did [this] right from the grasses, to the water quality, to the preservation of the creek bank, those are the real things I saw, that's what I'm working on right now. (Quoted in Bateman 2002, 5).

Community-based action incorporates locality-specific team building and tool building into communities that express an interest in working on riparian issues, allowing individuals and groups to set the direction and priority of their efforts. For example, in determining which reaches of a river could be inventoried to determine a health status, how neighbours can pool labour and resources to obtain new technology such as shared pumps for off-stream watering, or identifying sites they can use for demonstration purposes to show others the steps being taken locally to improve riparian zone management. Here, the role of Cows and Fish is to liaise, co-ordinate, guide and link people to each other and to technical/financial resources, as well as to assist with the formation of structured community groups where appropriate. All of these activities enable communities to take direct interest in and responsibility for learning about and managing their riparian zones. In describing the essence of Cows and Fish, a programme participant said: "Thinking about the Cows and Fish programme as a whole ... the key would definitely be the rancher grassroots involvement ... without the rancher community, I don't think there would be a Cows and Fish programme, I think we'd be in the same place we are in the States" (Quoted in Bateman 2002, 107).

The fifth element in the Cows and Fish Process involves ongoing monitoring of the ecological health of riparian zones, evaluation and adjustment. Grounded in adaptive management principles, a key monitoring tool is the riparian health inventory which, once demonstrated, allows a producer to assess the status of several ecological functions and identify an overall health category for a given riparian zone (Cows and Fish 2004c). Various management and grazing strategies can then be contemplated and the most appropriate implemented. Producers are encouraged to monitor their efforts, repeating the inventory and assessment as necessary to adjust their management strategy.

A formal evaluation of the Cows and Fish programme (Bateman 2001) examined the effectiveness of its awareness tools aimed at promoting the concept of ecological literacy, as well as the attitude-behaviour relationship that underlies it. Focus group participants gave the highest effectiveness ratings to awareness tools that were field-based, hands-on, personalised and specific to local landscapes (for example field events, programme staff, profile producers, and demonstration site tours). These tools provided opportunities for producers to observe riparian function and management options and exchange information about those topics in familiar surroundings. Maximising opportunities for family members and neighbours to share management information in a variety of interactive field settings was essential for learning. These tools, when delivered as part of a locally driven and appropriately paced process, had the most impact in developing ecological literacy, in particular when they provided flexibility for producers in choosing management options. These results reflected the community-based core values that frame the Cows and Fish programme.

Telephone interviews with 91 southern Alberta ranchers were conducted to assess the development of ecological literacy. The interviews were conceptualised around the Theory of Planned Behaviour (TPB) (Ajzen 1991), a psychological model that explains the attitude-behaviour relationship. TPB posits that a change in an individual's behaviour will not occur in the absence of an intent to modify behaviour. Intent, in turn, is underpinned by three distinct components: attitudes (which include knowledge), subjective norms and perceived behavioural control, components that mirror the elements of the Cows and Fish Process.

Interview results revealed a modest but statistically significant positive relationship between exposure to Cows and Fish awareness tools and, for example, greater knowledge about riparian function and the links between riparian zone health and water quality. Bateman also found a modest but statistically significant link between exposure to awareness tools and adoption of two sustainable grazing systems: time-controlled grazing, where livestock access to riparian zones is carefully manipulated through the growing season, and riparian pasturing, in which the riparian zone is defined as a distinct pasture unit, separate from upland areas. Of note was that the subjective norm component of TPB, the influence on decisions that an individual assigns to the preferences of significant others such as family members, friends and neighbours (Fishbein and Ajzen 1975), was also associated with the use of rotational grazing. This finding reinforced the group aspect of lifelong learning described by focus group participants when they defined the characteristics of awareness tools used by Cows and Fish that had the most impact on their attitudes and behaviour.

Conclusion

The seminal notion of B.F. Skinner's (1938) theory of operant conditioning was that learning leads to a change in behaviour. Lifelong learning about the causes and impact of environmental degradation and methods to avoid it should therefore lead to lifelong changes in the resource management techniques that are applied to the land. This chapter has shown that opportunities for lifelong learning, whether informal and unstructured as in the case of the Dobson Farm, or more structured community-based programmes such as Cows and Fish, have both led to a change in farmer and rancher behaviour. Both formal and informal modes of unstructured learning have clearly enhanced producers' understanding of the value of biotic diversity and ecological health on the farm and ranch. They have led to new and progressive attitudes towards the riparian zone and resulted in tangible changes in grazing land management. These changes in behaviour are clearly revealed on the agricultural landscape. In the case of Bob Dobson's farm, before and after monitoring photographs provide vivid *representations* of enhanced biodiversity over a period of years – an interval marked by a change in grazing land management practices (Dobson Farm 2004). In the case of Cows and Fish, the knowledge and attitudes of producers exposed to the 'Cows and Fish Process' have been examined in a formal evaluation and were shown to reflect the positive impact of this lifelong learning programme (Bateman 2001). Cows and Fish (2004d) demonstrates the Skinnerian 'change in behaviour' by *presentations*, most notably in the form of producer stories, vivid vignettes that show in narrative form how individuals and groups have learned about their environment, changed the way they manage their livestock, and identified personally with the positive and tangible outcomes in the appearance of Alberta's pastoral landscape. In both Dobson's personal philosophy and the more structured Cows and Fish Process, the message and the techniques are transmitted through an active social process that is embedded in the community and made vividly experiential by sharing field-based demonstrations that enable first-hand observation and a sense of trust that is shared among fellow producers. In both of these ways, lifelong learning is helping to cultivate a new cattle culture with a direct impact on pasture land management.

References

Ajzen, I. (1991). The theory of planned behavior, *Organizational Behavior and Human Decision Processes* 50, 179-211.

Bateman, N. G. (2001). *Evaluating the Alberta Riparian Habitat Management Program.* Unpublished M.A. Thesis. Lethbridge: School of Graduate Studies, The University of Lethbridge.

Bateman, N. G. (2002). *Evaluating the Effectiveness of the Cows and Fish Program: Evaluation summary, 2002.* Report prepared for Alberta Riparian Habitat Management Program. Lethbridge, Alberta.

Belsky, A. J., Matzke, A. and Uselman, S. (1999). Survey of livestock influences on stream and riparian ecosystems in the Western United States, *Journal of Soil and Water Conservation*, 54, 419-431.

Bradley, C. and Wallis C. (1996). *Prairie Ecosystem Management: An Alberta perspective.* Occasional Paper No. 2. Lethbridge, Alberta: Prairie Conservation Forum, Alberta Environmental Protection.

Breen, D. H. (1983). *The Canadian Prairie West and the Ranching Frontier: 1874-1924.* Toronto: University of Toronto Press.

Brotherton, I. (1989). Farmer participation in voluntary land diversion schemes: Some observations from theory, *Journal of Rural Studies,* 5, 299-304.

Brotherton, I. (1991). What limits participation in ESAs?, *Journal of Environmental Management,* 32, 241-249.

Canadian Cattlemen (1995). Awards recognize Canadian stewards of the land, *Canadian Cattlemen,* 58, 12-16.

Chilson, P. (1997). Tree huggers and damned cowboys: The rhetoric of an environmental conflict, *North American Review*, 282, 38–45.

Comer, S., Comer, E., Ekanem, S., Muhammad, S., Singh P. and Tegegne, F. (1999). Sustainable and conventional farmers: A comparison of socio-economic characteristics, attitude, and beliefs, *Journal of Sustainable Agriculture* 15, 29–45.

Commissioner of the Environment and Sustainable Development (2001). *Annual Report*, Chapter 1, Section 4. Ottawa: Office of the Auditor General of Canada. www.oag-bvg.gc.ca/domino/reports.nsf/html/c101sec4e.html (accessed October 2004).

Cows and Fish. (2004a). *About Cows and Fish,* Alberta Riparian Habitat Management Program. www.cowsandfish.org/about.html (accessed November 2004).

Cows and Fish. (2004b). *The Cows and Fish Process*, Alberta Riparian Habitat Management Program. www.cowsandfish.org/process.html (accessed November 2004).

Cows and Fish. (2004c). *Riparian Health Assessment*, Alberta Riparian Habitat Management Program. www.cowsandfish.org/health.html (accessed November 2004).

Cows and Fish. (2004d). *Community and Producer Stories*, Alberta Riparian Habitat Management Program. www.cowsandfish.org/stories.html (accessed November 2004).

Dobson Farm. (2003). *Environmental Stewardship Dobson's Grass Fed Beef Farm.* www.dobsonfarm.com/enviro.html (accessed October 2004).

Evans, S. (2001). Grazing the Grasslands: Exploring the conflicts of relationships and futures, *Prairie Forum,* 26, 67–84.

Fishbein, M. and Ajzen, I. (1975). *Belief, Attitude, Intention and Behavior: An introduction to theory and research.* Reading, Mass: Addison-Wesley.

Fitch, L. and Adams, B. W. (1998). Can cows and fish co-exist?, *Canadian Journal of Plant Science,* 78, 191–198.

Fitch, L., Adams, B. and O'Shaughnessy K. (2003). *Caring for the Green Zone: Riparian areas and grazing management.* Lethbridge, Alberta: Cows and Fish Program.

Fleischner, T. L. (1994). Ecological costs of livestock grazing in Western North America, *Conservation Biology,* 8, 629–644.

Jacobson S., Sieving K., Jones G., and Van Doorn, A. (2003). Assessment of farmer attitudes and behavioral intentions toward bird conservation on organic and conventional Florida farms, *Conservation Biology*, 17, 595–606.

Jordan, T. (1993). *North American Cattle-Ranching Frontiers: Origins, diffusion, and differentiation.* Albuquerque: University of New Mexico Press.

Morton, A. (1938). History of prairie settlement. In W. Mackintosh and W. Joerg (Eds.) *Canadian Frontiers of Settlement Vol. 2.* Toronto: MacMillan of Canada.

Rhodes, H., Leland, L., and Niven, B. (2002). Farmers, streams, information, and money: Does informing farmers about riparian management have any effect?, *Environmental Management,* 30, 665–77.

Skinner, B. F. (1938). *The Behavior of Organisms: An experimental analysis.* New York: Appleton-Century-Crofts.

Statistics Canada (2004). *Census of Agriculture 2001.* Ottawa: Statistics Canada.

Webster, M. (1994). Is beef getting a bum rap?, *Harrowsmith: Canada's Magazine of Country Living,* Number 113, 30–37.

Wilson, G. (1997). Factors influencing farmer participation in the Environmentally Sensitive Areas Scheme, *Journal of Environmental Management,* 50, 67–93.

13. ENVIRONMENTAL EDUCATION AND LIFELONG LEARNING: AWARENESS TO ACTION

MICHAEL WILLIAMS

Introduction

The terms environmental education and lifelong learning are both relatively recent additions to discussions about education. I would suggest that nowhere are they yet fully embedded in policies or in educational practice; they are still in the pioneering stage. Not surprisingly, they are often clouded in a rhetoric that impairs engagement with realities. Superficially, it would appear that environmental education, perceived as formal study in an educational institution or informally in vernacular adult settings, can be relatively easily discussed and that it could be quite unproblematic. However, even a brief consideration of the concept indicates that this is not the case.

To set the scene for the discussion that follows I put forward the following premises:

- The word environment means something different to each individual. This personal uniqueness has enormous implications in any consideration of what and how persons are motivated to learn and how organisations set about designing programmes for promoting environmental education.
- For environmental education in formal institutions – schools, colleges and universities – there is no universal agreement about where in curricula environmental education should appropriately be located.
- Given the vast array of knowledge, attitudes and skills that constitute environmental learning and the interplay between these three domains, it is not surprising that taste, or connoisseurship, figures prominently in the priorities given to environmental issues by theoreticians, policy makers, individual learners and individual educators.
- Just as the environment means different things to different individuals so, at various scales – local, regional, national and global – environmental education is spatially and culturally determined.

G. Humphrys and M. Williams (eds.), Presenting and Representing Environments, 197-213.

- Over their life span irrespective of their age, learners with access to the mass media obtain powerful environmental learning experiences that are not directed at any particular audience: as a context for learning, this environmental information content is a shared community resource.

These ingredients are mixed into a framework that serves as a scaffolding for the consideration of some of the central issues concerning environmental education and lifelong learning. I shall argue that how to present and represent the environment to learners at various stages in their lifetimes is always a difficult area for educators and educationists. The technical, pedagogical questions of how to teach and how to learn about the environment have often taken precedence, with questions about why one should teach about particular environmental issues and how to present unresolved and unclear environmental issues receiving less attention. This is more evident in post-school education than in school education, since in many societies the choices have been made outside the schools by agencies responsible for designing curricula. To facilitate the discussion I have drawn up a diagram (Figure 13.1).
In Figure 13.1 four inter-related aspects of lifelong learning have been highlighted:

- the stage of maturity or level of development of the learner;
- the mode of learning (pedagogy and andragogy);
- what is learned (knowledge and understanding, attitudes and skills);
- and the motivation that the learner brings to the learning content conceptualised at different scales.

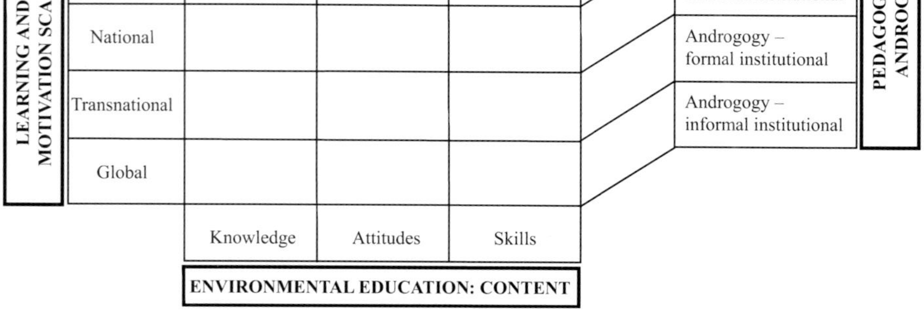

Figure 13.1 *A framework for environmental education and lifelong learning.*

These four aspects will be clarified first before discussing the distinction between informal and formal environmental education and then considering the arguments propounded for designing educational activities that lead to active environmentalism.

Lifelong Learning

Lifelong learning is one of those terms that suddenly surfaces in educational discourse. Clearly, it crops up in everyday conversations, carrying a self-evident truth that one never ceases to learn: day-by-day, minute-by-minute all of us learn something new about something. To that extent, we are all lifelong learners. However, the term has a more complex usage when employed by educationists and politicians. They use it as a term to guide policy and practice. Thus, it was much popularised by international organisations, particularly UNESCO, the European Union and the OECD in a number of documents published in the 1970s and 1980s (Field 2001). With the approach of the new millennium, it appeared in many countries across the globe as a guiding concept in the prefaces of official policy documents principally concerned with education and training, usually set in the context of industrial and commercial development.

We need to be clear that documents derived from international organisations must, in the context of national policies and practice, be treated cautiously. As Istance reminds us,

> They [international organisations] have been able to develop their programmes without close reference to the inconvenient factors of resource constraints, vested interests, bureaucracies, party politics, and the complexity of governance in the face of declining national state powers. Similarly, officials and politicians attending international gatherings are readier to endorse radical ideas when freed from everyday political constraints and untidy institutional realities. Lifelong learning is a concept that is much easier to agree on in broad principles than it is to implement in practice. It is by nature horizontal, stretching far beyond formal education and training systems, its stakeholders are diverse, and the relevant decision-making a complex mix of the public and private.
>
> (Istance 2003, 50)

Istance was writing about lifelong learning in general though much the same can be said of environmental education, as can be seen, as examples, in the much-quoted publications of the Belgrade Conference on Environmental Education (1975) (UNESCO 1976), the Tbilisi Intergovernmental Conference on Environmental Education (1977) (UNESCO 1978) and the Moscow Conference on Environmental Education and Training (1987) (UNESCO/UNEP 1987). It is important to keep his caveats in mind since many of the most challenging ideas and recommendations for practice in both lifelong learning and environmental education have emanated from international fora.

With regard to lifelong learning, from the literature, it is possible to trace over recent decades shifting emphases in the way it has been defined. Many commentators have distinguished between a liberal education view of lifelong learning, emphasising the self-fulfilment of individuals, and a vocational, performance and skills-based view. On the one hand, lifelong learning is perceived as the learning of the individual for his or her own purposes as these change over the life span. It is self-determined and self-regulated. Individuals are more or less conscious of their learning and may or may not seek in a deliberate and planned way to determine what they will learn, when, how and where. On the other hand, it has been argued that paying attention to lifelong learning is a governmental response to the perceived threat to national economies from the forces of globalisation, in particular increased competition in trade and commerce, leading to a demand for more sophisticated and flexible workforces. Here, the agenda for lifelong learning is perceived as a joint responsibility of employees and their employers. The assumption is that both groups share a common interest in productivity and efficiency and learning is defined as that what is required to meet the common interest during the working life of the employee.

What is important is to notice that an emphasis on liberal education leads into the territory occupied by the classical cultural component of traditional formal adult education in Western societies whereas the second emphasis leads into a vocational training culture. There is a third emphasis, the communitarian perspective, and this has been highlighted in the context of citizenship education. It is this latter emphasis that has particular relevance for environmental education. Using the work of Eder (1999) as a guide, Delanty, writing in the context of citizenship as a learning process, argues:

> It is important to see the learning component not just in individual terms but also as a medium of social construction by which individual learning becomes translated and co-ordinated into collective learning and ultimately becomes realized in social institutions. What is decisive in this is collective learning, which cannot be reduced to individual learning … What is learnt on the collective level is very different from what is learned on the individual level. (Delanty 2003, 601).

There is a link between the collective learning described by Delanty and the distinction between *human capital* and *social capital* commonly encountered in contemporary discussions about lifelong learning. These terms have been defined in a recent report on national well-being published by the OECD. Thus, *human capital* is defined as, "The knowledge, skills, competencies and attributes embodied in individuals that facilitate the creation of personal, social and economic well-being" (OECD 2001, 18). The focus is on individuals and their well being. *Social capital* is defined as "networks together with shared norms, values and understandings that facilitate co-operation within or among groups" (OECD 2001, 41). Clearly, this definition emphasises interpersonal aspects and links into Delanty's definition. It would be easy to argue that human capital is the concern of formal schooling but Healy warns that

> we should not imagine that human capital equals formal schooling or years of education completed. Learning is much too rich and varied and acquired in so many different settings throughout life that the 20–25% of total waking time spent by young people in formal education from age 5 to 15 is only part of the picture – even though an extremely important one. (Healy 2002, 5)

It is worth pausing to reflect on the particular relevance for environmental education of this distinction between human and social capital. At whatever scale one engages with environmental issues, the individual wishing to move from some general state of awareness to a more active involvement increasingly has to interact with other individuals. Individuals acting alone can go only a limited way towards making sense of any environmental issue. They need to improve their learning through interacting with others; others who may reinforce their understanding or extend their understanding as well as offering the reassurance, encouragement and praise to carry them from awareness through understanding to some form of action. Fundamental to this learning is the importance that must be given to the distinction between what any organisation or institution wishes people to learn and what the learner is disposed to learn. Environmental education has a set of meanings for the educator and another set for the learner and these change over time, as the individual matures.

Stages of Maturity

It is obvious that the environmental learning profile of any individual or group has to take age into account. With increasing age, individuals have greater opportunities to acquire environmental knowledge, attitudes and skills. With increasing experience their environmental priorities will also change. The innocence of the young merges into the ideological awakening of the adolescent and the sharper focusing on particular aspects of the environment and environmental issues in middle age. Age brings concerns that are of immediate, personal interest to individuals and communities and these concerns may be located in one's own backyard, in one's neighbourhood or in larger areas. There is no shortage of local issues or of single interest active groups as well as the parallel wide range of national and international concerns – from global warming, the destruction of the rainforests, the threats to Antarctica, the potential of GM foods, to the plight of numerous species of flora and fauna. Increasing maturity leads to a consideration of how, in any culture, however defined, environmental awareness and environmental action vary over the lifespan. This can be explored through such questions as: What are individuals capable of learning and what is the extent of the variations of this capability over time? What are the learning context (where people learn) variations over time? What differences are there in the formal and informal institutional and non-institutional places where individuals learn? How and by whom, where, when and why, and with what impact are different presentations and representations of the environment communicated to learners in varying contexts of time and space? Even a superficial consideration of learning opportunities and learning possibilities reveals the vast differences between the learning of the house-bound infant and the learning of the pensioner.

In the context of lifelong learning it is important to recognise not only shifts in opportunities, interests and enthusiasms with age but also the way persons learn at various stages in their development. A useful way to approach this is to consider the differences between pedagogy and andragogy.

Pedagogy and Andragogy

The use of the terms pedagogy and andragogy has served to distinguish between the modes of teaching and learning employed commonly in schools and those used in post-school contexts. It is the synthesis of the characteristics of both of these that lies at the heart of lifelong learning. Schools need to pay increasing attention to the way adults learn in various contexts, and adult educators need to pay increasing attention to the way school students learn in various contexts.

What is equally important to the formal, organised and structured learning that takes place in schools, seen as a temporal stage of development, is the learning in which school students engage during out-of-school hours. The school structure is usually based on a linear progression of developmental stages demarcated by students' learning age. Relatively restricted, simple and controlled school learning experiences contrast with the rich, complex and uncontrolled diversity of the out-of-school learning experiences of school students. This gives the pedagogy–andragogy distinction a special dimension: it brings to the fore a horizontal quality, the supplementation of school learning by extra-curricular learning beyond the school site, that schools must increasingly recognise. This argument suggests that schools will improve as they engage in successfully exploring, clarifying and implementing appropriate modes of learning drawn from andragogy.

The word andragogy became familiar to educators following the publication of Knowles' pioneering book on the modern practice of adult education (1970, revised 1980). He sought to emphasise that pedagogy and andragogy were two models based on assumptions about learners rather than a crude distinction between pedagogy as a form of learning characteristic of schooling and andragogy as a form of learning characteristic of the education of adults. Thus he argues "…the models are probably most useful when seen not as dichotomous but rather as two ends of a spectrum, with a realistic assumption in a given situation falling in between the two ends" (Knowles 1980, 63).

Knowles used four assumptions as the basis for his analysis: the concept of the learner; the role of the learners' experience; readiness to learn; and the orientation to learning. Whereas in pedagogy the learner is conceptualised as being dependent, in andragogy the assumption is the learner is increasingly self-directed. In pedagogy the role of the learner's experience is relatively less important while in andragogy the learner's experience is a highly valued resource. Knowles argued that in pedagogy learners were ready to learn whatever some higher authority claimed they needed to learn provided there was strong pressure on them to learn. On the other hand, in andragogy the assumption is that the learners learned when they were faced with a real life task or problem to resolve. Finally, (referring to the orientation to learning) in pedagogy he stated that learners saw education as "a process of acquiring subject-matter content, most of which they understand will be useful only at a later time in life" (Knowles 1980, 44). In andragogy, education was seen by the learners as "a process of developing increased competence to achieve their full potential in life" (Knowles 1980, 44).

This distinction between pedagogy and andragogy serves to draw our attention to differences in the learning process. It alludes to the importance of motivational differences between young and adult learners. A concern with motivation directs us to the search for understanding why different individuals become more aware, more interested and more demonstrably active with regard to environmental issues. Within the school context, this has particular relevance for the way children are encouraged to learn about the environment. School students, like students in higher education, are enthusiastic about some conventional subjects and shy away from others. They may be positively disposed towards geography but not to history or science or mathematics or the creative arts. In some stages of education and in some countries the distinction between subjects may be blurred as new forms of curriculum organisation are introduced. The structural arrangement of knowledge is important for student motivation, student progression and student achievement. The subject structure underpins the way the environment is defined in schools: it prefaces what counts as worthwhile knowledge – what is worthy of formal presentation – and determines the way selected aspects of the environment are taught and learned. The way the environment is presented and represented to young learners conditions many of the routes that are available for learners to construct their own environmental learning.

It is but a short step from motivation to self-efficacy. In essence, motivation refers to a disposition to express concern. Self-efficacy refers to the sense that one can do something about the issue for which one has expressed concern. As Bandura wrote,

> Unless people believe they can produce desired effects by their actions, they have little incentive to act. Efficacy belief, therefore, is a major basis of action. People guide their lives by their beliefs of personal efficacy. Perceived self-efficacy refers to beliefs in one's capabilities to organize and execute the courses of action required to produce given attainments.
> (Bandura 1997, 2–3)

If action is a desired product of formal environmental education, then facilitating improved self-efficacy must be taken into account. Learners must be presented with environmental contexts in which they can clearly anticipate making a difference. Such action may be at the very local level of the school site and may be limited to such activities as litter removal, electricity saving or maintaining a school garden. It may also be at a local community level where interested adults seek to conserve particular resources or develop resources in an environmentally sustainable way.

While acknowledging, in passing, the strength of meaning implied in the term used at the start of this paragraph, action as a 'desired product', we can pause to reflect on the kind of experiences that are likely to impact powerfully on learners at various stages in their lifecycles. These 'powerful learning experiences' (a term that has been familiarised through the work of the US-originated Accelerated Schools Project, (Finnan, Schnepel and Anderson 2003)) can be planned, structured and provided in formal institutions but they can just as readily be gained from vernacular experience. In school-based environmental learning, there is an assumption, at least in England and Wales, that powerful learning experiences can be generated through conventional school subjects, particularly science and geography. This is a restricted view, denying the multi-faceted nature of the environment and restricting the forms of environmental presentation and representation. Learners, of whatever age, may be powerfully influenced by what they see, hear and feel as they go about their daily business: they learn through observation of the environment at first hand as well as what they encounter from a range of media.

Environmental Education: Awareness to Action

Environment is one of those portmanteau words that cover a wide range of meanings and interpretations. As with most other curricular areas, environmental education can easily be reduced to the simply stated knowledge and understanding, attitudes and skills. This, however, does not take us very far. It is what constitutes environmental knowledge, environmental attitudes and environmental skills that has to be addressed. The response is either to draw up taxonomies of aims and objectives as a step towards the selection of what will be taught or to identify bodies of knowledge that will accommodate what might broadly be termed environmental concerns. The crucial issue is that there has to be selection and any selection will bring into play a broad range of criteria and influential forces. As, Jelin has argued:

> What different actors define as 'environmental', the meanings attached to the concept, the implicit concept of agency – all these have a bearing on the paradigm within which 'environmental concerns' are interpreted in turn; they are set in the context of a broader understanding of the social order, of the meaning of entitlements (and ultimately of democracy), and of the relationship between humanity and nature. (Jelin 2000, 48)

It is not simply a matter of deciding whether to focus on particular aspects of the 'physical' or 'natural' environment, or indeed on aspects that embrace both 'natural' environments and environments created by human agency. There are also other questions related to the importance of environmental issues for learners at their stage of learning in the places in which they live. Place, in this context can be local, regional, global and virtual. Other questions relate to the availability of resources and the quality of those resources available to support teaching and learning. These include access to competent and appropriately qualified teachers as well as such materials as textbooks and teaching aids.

The principle that irrespective of age everyone should be, at least, environmentally aware underlies many of the arguments that are propounded by those who are keen to include environmental education in the curriculum for school children and to advocate environment-related public participation. It is one thing to advocate the inclusion of environmental education into the school curriculum and quite another to incorporate it fully in any meaningful way. There are many obstacles including the greater priority given by curriculum policy makers to other issues, such as literacy and numeracy, science education, citizenship education, computer education, health education (including anti-smoking education, drugs education, and sex education) and, in some countries, like Wales, to the national language and the national dimension.

In addition there are difficulties surrounding the environmental line that should be taken in schools. Environmental educators cannot agree about what should be taught, to whom, when, by whom and where in the curriculum (See Lee and Williams 2001). The arguments run from the inclusion of environmental awareness in science topics, where the emphasis may be on observation and experiment, through to the advocacy of education for sustainability that has a strong moral character and extends to the consideration of the need for societal, political and social transformation. The more sophisticated the arguments, the more difficult it is to find a way of developing environmental education in schools. For any approach to environmental education to be successful there has to be strong support from school management, from teachers, from parents and the local community, as well as from the school students. A national policy of compulsory environmental education can only be successful if resources, human and material, can be put in place. Given the pressure on finite curriculum time, the addition of a new field like environmental education can only be achieved by the removal of subjects, themes or topics that already have strong claims to curriculum space.

The most pressing case has recently been made for all school students to gain some understanding of sustainability and sustainable development. Both of these terms have proven difficult to accommodate. Just how complicated and challenging this task is can be seen in a recently published paper in a UNESCO publication (Ospina 2000). The author seeks to define sustainable development in a list of statements, from which I have chosen the following as illustrative of the gist of the definition:

- Sustainable development involves the natural sciences and economics, but it is primarily a matter of culture. It is connected with *values* people cherish and with the ways in which they perceive their relationships with others.
- Sustainable development occurs when we acknowledge the relationship between human needs and the natural environment. The interdependence of humans and the environment necessitates a refusal of the obsessive pursuit of any single development or environmental objective to the detriment of others. The environment cannot be protected in a way that leaves half of humanity in poverty.

- Linking social, economic and environmental concerns is a crucial aspect of sustainable development. Creating such links demands a deeper, more ambitious way of thinking about education, one that retains a commitment to critical analysis while fostering creativity and innovation. In short, it demands an ethic and value system sensitive to the value of cultural identity and multicultural dialogue (Ospina 2000, 33).

Further, Ospina (2000) presents a list of seventeen goals to achieve sustainable development, that includes the following:

- To encourage lifelong learning, starting at the beginning of life and grounded in life – one based on a passion for a radical transformation of society and a change in the moral character of society.
- To advance new conceptions rooted both in traditional scientific rationality and in popular beliefs and consciousness, drawing on these as a source of humane understanding and a pointer to collective wisdom.
- To promote a culture of citizenship and give value to social actors (such as, non-governmental organizations and other sub-groups).
- To reach a stage in which the possibility of change and the real desire for change are accompanied by a concerted, active participation in change, at the appropriate time, in favour of a sustainable future for all.
- To develop to the maximum the potential of all human beings throughout their lives so that they can achieve self-fulfilment and full self-expression with the collective achievement of a viable future (Ospina 2000, 34–35).

Such lists are loaded with principles derived from ideological positions that create tensions in debates about the spirit and purpose of environmental education. Fien (1993, 40) has lucidly demarcated the variety of ideological positions in a matrix that has three educational ideologies on one axis: vocational/neo-classical, liberal-progressive and socially-critical; and, on the other, two environmental ideologies technocentric and ecocentric. Associated with these there is a rich array of '-isms' – gaianism, utopianism, managerialism, ecosocialism, and communalism together with environmental education about, through and for the environment. In reading these, we have to reflect on what sense can be made of them by busy teachers who are encouraged to engage in environmental education, while often lacking the necessary academic and professional competencies. It is interesting to read arguments, derived from critical theory, that define teachers as "critically reflective, transformative intellectuals" (Huckle 1995, 15). This is some way from the simplistic 'environmental trinity' that is commonly encountered in the environmental education literature: education *about* the environment; education *through* the environment; and education *for* the environment.

Such discussions yield a variety of substantive categories. The criteria used for the selection of these as being appropriate for the environmental education of children and adults are grounded in ideological preferences, translated by Gough, Walker and Scott

(2001) into theories of action. Their preferences can be translated into a number of broad approaches, each of which highlights a different mode of presenting the environment.

- *Classical humanism:* where the emphasis is upon presenting environmental knowledge for its own sake and where such knowledge is seen within conventional school subjects, and where values exploration and the acquisition of environmentally related skills may be a secondary consideration.
- *Scientific:* where the emphasis is upon scientific knowledge acquisition through teacher controlled observation, investigations and experiments.
- *Recreational:* where the environment is presented as a stimulus and a location for such recreational activities as photography, painting, writing and walking.
- *Vocational:* where the environment is presented as a source of potential employment and where skills acquisition dominates.
- *Citizenship:* where the environment is presented as a contested area where communitarian activities come to the fore.
- *Ethical:* where the environment is presented as a source of values awareness and clarification and where such issues as intra-generational and inter-generational justice are explored.
- *Economic:* where the environment is presented in terms of environmental audits, environmental accounting and accountability, environmental justice and industrial and commercial greening.
- *Legal:* where the environment is presented as an area of contested rights and where environmental values are translated into the interests of individuals, corporate groups, national governments and international agencies.

How these are translated into educational policy can be traced in educational changes. Thus, in England and Wales, a National Curriculum was introduced into the schools by the Education Reform Act of 1988. This was defined by subject titles and the content of each subject was specified as a list of statutory requirements. Schools also received non-statutory guidance and environmental education was part of this. Environmental education was perceived as a cross-curricular theme that could be included in various subjects as and when particular schools judged it to be appropriate (Curriculum Council for Wales 1992). It was for schools to decide the approach, the priority to be given to particular issues and the amount of resources to allocate to it. Since 1988 the National Curriculum has been subject to various changes and one of the results has been the squeezing out of non-essential studies and a greater focus on a limited range of subjects, particularly science, mathematics and language. In these fields targets have been set and a range of monitoring activities introduced to ensure that schools keep these targets to the fore, for example, examinations and tests, school inspection and league tables. Environmental education does not fit readily into this culture. This is not to say that enterprising schools have failed to introduce students to many worthwhile environmental learning opportunities though many of these are extra-curricular, taking the form of competitions and projects that may be strongly supported by various community agencies.

School education of this kind and post-school education are linked by the influence brought to bear by various agencies in non-institutional settings. Thus young people and adults share access to television programmes and increasingly to the Internet. Both groups have access to the press and to both news coverage and advertising in newspapers and magazines. Both groups are targeted by a wide range of environmental organisations that seek members and support for a wide range of educational and political activities. For some environmental organisations the conventional educational role is paramount while for others there is greater interest in political activism including lobbying of local, national and international organisations, high profile campaigns, the collection of signatures on petitions and direct action in the form of protest marches and demonstrations.

For both adults and children it is important to recognise the limited effects of the media in achieving behavioural change, even of television as a popular medium. We have only to reflect on the impact made by the AIDS campaign, the anti-drugs, anti-alcohol and anti-smoking television and school-based campaigns to realise that despite their sophistication and high costs they seem to have disappointing results.

In addition to the media and specialist organisations there are also those organisations that cater specifically for adult learners in a formal way. Here we are thinking about local authorities and voluntary organisations that make institutional arrangements for adult education. Also included are the community education providers in universities and colleges. In British formal institutions, especially in universities, there is a continuing strong influence of what might be described as the classical adult education curriculum – social history, sociology, political economy, English literature. These have long attracted a particular group of learners interested as individuals as well as members of groups and communities – often disadvantaged. Alongside this is the inevitable array of information technology courses and workshops. Outside the universities formal adult education provision is often driven by funding considerations – they who hold the purse strings determine what is provided when, where and how. The environment may figure in this provision, though rarely prominently.

While the National Curriculum is the basis for what goes on in English and Welsh schools, Agenda 21 provides the basis for much community environmental education:

> Local authorities construct, operate and maintain economic, social and environmental infrastructure, oversee planning processes, establish local environmental policies and regulations, and assist in implementing national and subnational environmental policies. As the level of governance closest to the people they play a vital role in educating, mobilizing and responding to the public to promote sustainable development.
>
> (United Nations 1993, Agenda 21, Chapter 28).

For educationists there is a special interest in finding out why some individuals and groups are more environmentally active than others. Why is it that some individuals are particularly concerned about particular environmental phenomena?

Why is it that some individuals are attracted more than others to environmental direct action and to lead direct action? It is clear that, despite a common formal education, some individuals emerge as environmental activists who engage in activities that clearly mark them out as activists. Eco-warriors are the clear examples of this. Persons who chain themselves to road making equipment or construct tree shelters or dig themselves into tunnels to prevent environmental despoliation are overt, active environmentalists, individuals or members of groups who seek to make a difference. There is no doubting their high level of motivation and high sense of self-efficacy. Where are the roots of this activity? What are the pathways that lead to such activity? To what extent are these roots and pathways different for people from different backgrounds – gender, race, social class, intelligence, age, type of formal education? The very confused and heterogeneous nature of the 'environment movement' at whatever scale one cares to focus renders the search for answers to these questions problematic. To see this, we have only to think of the wide range of environmental issues that touch our individual lives and to consider those that have led us to seek more information, to join single interest or multiple interest organisations, to engage in various activities and to join local, national, international groups and networks. We can also ask why there are environmental issues and activities in which we have no interest and which we simply ignore.

We are all aware of many organisations that constitute the environmental movement: of the green parties, of scientists with strong ecological interests, of media personalities and groups who seek to influence public opinion, of corporate businesses that advertise environmentally sympathetic products, and of local and national governmental agencies that are environmentally active (Jelin 2000). It would be easy to assume that pro-environmental interests are unchallenged but, as Rowell (1996) so clearly documented, it would be foolish to ignore the strength of the anti-environmentalists. He documents the development of the Wise Use movement in the US that brings together 'a coalition of ranchers, miners, loggers, farmers, fishermen, trappers, hunters, off-road vehicle users, property rights advocates, industry associations, corporate front groups, and right-wing activists…' (Rowell 1996, 14).

The educational activities of these right-wing groups has been studied by Hill in the US and he wrote:

> The rightist movement not only produces different knowledge – but also produces knowledge differently from non-conservatives. It places the *market* as the central axis for making sense, as opposed to *ecological principles*. Adult educators of this alternative environmentalism employ various mechanisms to shape popular opinion and behaviour. Reactionary environmental knowledge has its own rules and institutions of production, sites of learning, pedagogical tools/methodologies, and dissemination systems. (Hill 2002, 187)

He goes on to discuss the educational tactics used by the anti-environmentalists. They comprise a long list that he divides into methods of disempowerment and methods of empowerment. Included in methods of *disempowerment* are marginalising practices in which pro-environmentalists are presented as being 'on the fringe' and deception-by-omission where publications are produced that omit reputable scientific information. Some evidence of environmental problems are denied and, by a process of 'greenwashing', publications from official and other sources are presented as value-free despite being value-laden. Turning to methods of *empowerment*, Hill (2002, 190). refers to the embracing of conflict, the connecting of anti-environmentalist activity to broader social discontent and the construction of group self-identity. Groups within the Wise Use movement had adopted patriotic names including People for the USA; the Heritage Foundation; Alliance for America; and the American Enterprise Institute. Such organisations, Hill claims, seek "creative outlets for expressing ideas and beliefs, and link their messages to the everyday lived experience of their learners" (Hill 2002, 191) and engage in power networking. In particular they seek to influence women's groups, non-governmental organisations and academics.

It is interesting to locate these formal school curriculum approaches and grassroots environmental movements within the agendas set by international and national governmental bodies. In this context it is interesting to read two recently published papers set alongside each other (Kelsey 2003; Rose and Bridgewater 2003).

Rose and Bridgewater focus on UNESCO, reminding us that this organisation is the task manager for Chapters 35 (Science) and 36 (Promoting education, public awareness and training) of Agenda 21 produced by the United Nations Conference on Environment and Development held in Rio de Janeiro in 1992. They proceed to show the educational role that UNESCO plays in implementing four multilateral environmental agreements: the Convention on Biological Diversity, the United Nations Framework Convention on Climate Change, the United Nations Convention to Combat Desertification and the Ramsar Convention on Wetlands. In particular they emphasise the importance of outreach programmes for which they use the term "communication, education and public awareness" (Rose and Bridgewater 2003, 264). They argue that governments must: introduce measures that organise the "educational process to include the entire range of potential resources in society"; "develop new programmes, methodologies and approaches...that help citizens to make appropriate decisions"; and produce "relevant and forward-looking scientific and technological information" and develop "effective dissemination and communication mechanisms for improving access to such information by civic society" (Rose and Bridgewater 2003, 271).

Kelsey (2003) uses discourse analysis to investigate three multilateral environmental agreements: the Convention on Biological Diversity, the Convention on Access to Information, Public Participation in Decision-Making and Access to Justice in Environmental Matters, and the Draft Framework for Public Participation in the Commission for Environmental Cooperation's Activities. She argues that these three agreements provide excellent examples of administrative rationalism. This "seeks to organise scientific and technical expertise into a bureaucratic hierarchy in the service of the state" (Kelsey 2003, 416). Her analysis suggests that, "Though unstated, the goal is not simply public participation, but rather participation that returns power to government and ensures public compliance to government-determined procedures" (Kelsey 2003, 418). She is critical of a discourse that sees environmental knowledge as the means of resolving environmental problems and public participation as operating within state-defined conceptions of appropriateness. Further, an emphasis on public information dissemination, especially via the mass media and electronic forms of communication, denies the importance of knowledge construction by the recipient. The assumption is "that learning is a passive event, that knowledge can be acquired with little or no effort on the part of the learner" (Kelsey 2003, 422).

In the literature of school-based environmental education there is frequent advocacy of environmental activism that embraces constructivist theories of learning. However, as in public participation, as envisaged in international conventions, such student activism is perceived as being socially appropriate, socially desirable and socially approved. What is inappropriate for students may be considered to be appropriate for adults. Much depends on how action is presented and represented. Here lies the education conundrum: the environment is presented and represented in various culturally determined ways and the responses to these presentations and representations are also presented and represented in various culturally determined ways. Bridging the environmental learning of the young with that of the mature is the task of those who seek to promote environmental lifelong learning. In this bridging, the reconciliation of learning for the individual and learning for the community has to be achieved.

Conclusion

Environmental education takes many forms varying from the vernacular experience of children and adults through to formal provision in school and post-school curricula. It is defined in many ways and priorities vary from place to place, from culture to culture, from teacher to teacher and from learner to learner. While environmentalists are anxious to ensure that more and more people are increasingly aware of environmental issues, they differ in their ideologies and the behaviour that stems from them. What is clear is that, despite the attention paid to the environment by international and national organisations, environmental education, seen as a holistic, inter-disciplinary study that takes the learner from awareness through to action, is generally marginalised in the formal curricula of educational institutions and, in the informal curricula, there are mixed messages that may bore or confuse some and stimulate others to direct action.

These mixed messages are the products of several significant forces. As has been indicated, there are difficulties simply in the language used to discuss the environment. The environment means different things to different stakeholders: for some it is the subject for what appears, on the surface at least, to be detached scientific investigation, for others it a source of complex problems, for others it is a stimulus to aesthetic creation and a spur to rich imagination, and for yet others it is a focus for promoting social transformation. The shift in schools from nature study to environmental studies to environmental education to education for sustainability to education for sustainable development has confused the educational scene. Teaching and learning about the environment at whatever stage in the lifespan is complicated and immersed in difficulties associated with ideological positions, curricular cultures, resource availability, community stakeholders and political pressures. Given the complexity, it is not surprising to find that, despite the strength of environmental voices, teaching and learning about the environment in a holistic and inter-disciplinary way in formal institutions is generally marginalised. This marginalisation contrasts with the powerful learning experiences gained by those learners of all ages who have access to the mass media. However, the mass media have their own ideologies, environmental enthusiasms and interests. The environment as a subject for entertainment and carefully selected campaigns conflicts with the earnestness of formal educational endeavours. Teachers, as reflective recipients of the media, are being challenged to make sense of the environment, sustainability and sustainable development and as adult learners they derive some of their information from their academic studies but a great deal more from a rich variety of media experiences. They have the enormous task of choosing appropriate definitions of environment for themselves before presenting these to their classes, of determining how they will present and represent the environment and determining how far they wish their learners to move from environmental knowledge to environmental action.

References

Bandura, A. (1997). *Self-efficacy: The exercise of control.* New York: W. H. Freeman and Company.

Curriculum Council for Wales (1992). *Environmental Education: A framework for the development of a cross-curricular theme in Wales: Advisory Paper 17*, Cardiff: Curriculum Council for Wales.

Delanty, G. (2003). Citizenship as a learning process: Disciplinary citizenship versus cultural citizenship, *International Journal of Lifelong Education*, 22, 597–605.

Eder, K. (1999). Societies learn and yet the world is hard to change, *European Journal of Social Theory*, 2, 195–215.

Field, J. (2001). Lifelong education, *International Journal for Lifelong Education*, 20, 3–15.

Fien, J. (1993). *Education for the Environment: Critical curriculum theorizing and environmental education.* Melbourne: Deakin University Press.

Finnan, C., Schnepel, K. C. and Anderson, L. W. (2003). Powerful learning environments: The critical link between school and classroom cultures, *Journal of Education for Students Placed at Risk*, 9, 391–418.

Gough, S., Walker, K. and Scott, W. (2001). Lifelong learning: Towards a theory of practice for formal and non-formal environmental education and training, *Canadian Journal of Environmental Education*, 6, 178–196.

Healy, T. (2002). Social capital and lifelong learning: Issues for public policy, *Lifelong Learning in Europe*, VII, 4–9.

Hill, R. J. (2002). Pulling up grassroots: A study of the right-wing "popular" adult environmental education movement in the United States, *Studies in Continuing Education*, 24, 180–203.

Huckle, J. (1995). *Reaching Out: Professional development for teachers, Part 1 (Introduction and Initial Perspectives).* Godalming: World Wide Fund for Nature (WWF).

Istance, D. (2003). Lifelong learning and citizenship: international perspectives. In M. Williams and G.

Humphrys (Eds.) *Citizenship Education and Lifelong Learning: Power and place*. New York: Nova Science Publishers, Inc.

Jelin, E. (2000). Towards a global environmental citizenship?, *Citizenship Studies,* 4, 47–63.

Kelsey, E. (2003). Constructing the public: Implications of the discourse of international environmental agreements on conceptions of education and public participation, *Environmental Education Research*, 9, 403–427.

Knowles, M. (1970, revised 1980). *The Modern Practice of Adult Education: From pedagogy to andragogy.* Englewood Cliffs, NJ: Cambridge Adult Education, Prentice Hall Regents.

Lee, C. K. J. and Williams, M. (2001). Researching environmental education in the school curriculum: An introduction for students and teacher researchers, *International Research in Geographical and Environmental Education*, 10, 216–244.

OECD (2001). *The Well-Being of Nations: The role of human and social capital*. Paris: OECD.

Ospina. G. L. (2000). Education for sustainable development: A local and international challenge, *Prospects*, XXX, 31–40.

Rose, O. H. and Bridgewater, P. (2003). New approaches needed to environmental education and public awareness, *Prospects*, XXXIII, 263.

Rowell, A. (1996). *Green Backlash: Global subversion of the Environment Movement*. London: Routledge.

UNESCO (1976). *Final Report of the International Workshop on Environmental Education, Belgrade (13–22 October, 1975).* Document ED-76/WS/97, Paris: UNESCO.

UNESCO (1978). *Final Report of the Intergovernmental Conference on Environmental Education Tbilisi*. Document Ed/MD/49, Paris: UNESCO.

UNESCO/UNEP (1987). *International Action Strategy for Environmental Education and Training for the 1990s*. Moscow: UNEP.

United Nations (1993). *Agenda 21: A Programme of Action for Sustainable Development: Rio Declaration on environment and development: The final text of Agreements negotiated at the United Nations Conference on Education and Development* (UNCED), 3–14 June, Rio de Janeiro. New York: United Nations agreement signed at the 1992 Rio Summit.

INDEX

acid rain 3, 155
actor-network theory 172
Agenda 21 97, 208
agro-environmental programmes 184
androgogy 198–204
anthropology 47, 71, 137
anthropomorphism 145
anti-foundationalism 171
arcadia 10
archaeology 47
Area of Outstanding Beauty 66
art 63, 71, 137–52, 167–80
arts 5–7
astronomy 170
author's gloss 80

Bathurst Mandate 168–9, 174, 176–7
bear-human relationships 38
behavioural change 10
bilingualism 153–66
biodiversity 20, 66, 153, 182, 193
biodiversity ethic 189
biological endowment 77
biological science 47
biologists 34, 39
biology 44, 171

campaigns
  anti-fishing 20
  anti-logging 20
  anti-mining 20
  anti-oil drilling 20
  anti-sealing 20
  anti-whaling 20
  language protection 162–13
  market boycott 25
  save communities 156
capital
  human 111, 200
  natural 12, 105,106, 111–15
  produced 111
  social 90, 111, 114, 200
cattle grazing 181–95
catastrophe theory 8
chaos theory 8
citizenship education 200, 205
classical humanism 207
Claude glass 140
climate change 10, 47, 66, 121–34
cognitive construction 75, 77–9
cognitive literature 78
cognitive skills 79
communalism 206
community defence 154
community empowerment 154
conservation biology 38
consumerism 63
Cows and Fish Programme 190–93
cross-cultural
  borrowing 9, 10
  contact 10
  context 38, 47
  issues 9–11
  setting 34
  studies 10
cross-disciplinary approaches 6
cultural
  attrition 153–6
  benefits 19
  clash 13

communication 46
conflict 9, 10
contact 10
conventions 138
gap 170
hierarchy 9
identity 11, 167–80
mythologising 10
sharing 10
studies 73
turn 10

decision process 35, 45, 46
deindustrialisation 19
democratic elitism 91
desertification 155
despoliation 95, 153
devolution 33
discourse analysis 3

Earth Art 146
Earth Summit 121
eclecticism 8, 18
ecocentrism 8, 31, 206
ecological
approaches 70
argument 157
health 192
history 57–60
integrity 46
knowledge 40, 46, 47, 172
literacy 190, 192
methods 47
modification 57
nexus 58
perspective 157
reality 173
revolution 59, 140
sciences 38
studies 74
systems 42, 44
texts 85
values 17, 23, 155
ecosocialism 206
ecozone 37
ecosystem 33, 35, 38, 43, 46, 47, 106
elitism 8, 90, 91, 138, 140, 158
Enlightenment 8
Enlightenment humanism 155
Enterprise Zone 97
environmental
accounts 12
aesthetics 137
action 11, 14, 197–213
assessment 96
attitudes 13, 14
benefits 13
challenges 13, 14
change 89, 90, 122
conflicts 33
education 13, 185, 197–213
management 89, 110
policy 11
politics 99
risk assessment 4
science 91
standards 1
sustainability 17
environmentalism 18
epistemology 172, 173
ethical
commitment 184
perspective 5
ethics 5, 7, 35, 48
ethnography 47
ethno-regionalism 156
Europe of the Regions 158
European Community 96
European Union 66, 157, 158, 199
evaluation 46, 62, 192, 193
exploration 69–87
exploration
literature 69–87
narratives 12, 72–87
texts 75–84

feminism 11, 144, 145
film 64, 65, 140
focus groups 34
Fordism 19, 20
forestry 10, 11, 17–32, 57, 106, 129
fossil fuels 126, 131–4

Gaia thesis 154, 158
Gaianism 206
geography 5–7, 19, 71–3, 99, 137
cultural 70, 173
perception 77
global economy 18, 19
global warming 122–34, 153
globalisation 10, 20, 153, 162, 175, 200
GM foods 10, 201
governance 8, 33, 90, 102, 109–11, 118, 153, 164–5
greenhouse effect 122, 133, 157
greenhouse gases 126

Greenpeace 17, 21–5, 27

habitat 10, 66, 109, 112, 123, 157
hermeneutic
  approaches 71
  ideas 73
humanities 5, 6, 7, 10

industrialisation 159
industrialism 64
industrial revolution 59
information and communications technology 19
information economy 185–7
Internet 1, 12, 101, 123
interviews 34, 40, 123, 192
Inuit
  art 167–80
  identity 11, 167–80
  values 169
irrigation 108–18

knowledge
  co-production 48
  dissemination 81
  ecological 40, 46, 47, 172
  integration 35, 40, 41
  local 40
  public 83
  sets 46
Kyoto protocol 121, 127

labour relations 25
Land Art 146–7
Landscape Scenery Model 141
learning
  double-loop 45, 49
  group 157
  lifelong 14, 181–95, 197–213
  passive 14
  single-loop 45
  social 36, 44, 45
liberal democracy 90–92, 169
liberal democratic government 12, 90
liberal education 200
liberal humanism 171
literary
  criticism 73, 84
  studies 73–5
  theory 70, 71
lobbying 19
localism 156

majority cultures 13
managerialism 206
market forces 19
media 1,2, 13, 60, 121–36
medium-size enterprises 30
metanarratives 8
metaphysics 172
minority
  cultures 13, 153–66
  languages 157
modernism 8
myth 10, 38–9, 47, 49, 60–61, 149, 157, 172
moral philosophy 154
motivation 198, 200
multiculturalism 9, 13, 153
multi-national corporations 20

nationalism 155–8
nationhood 69
negative artistic critique 13, 137–51
news-as-entertainment 3
newspapers 1, 77, 121–34
non-governmental organisations 4, 9, 10, 17–33, 119, 209, 210
normal science 35, 39

OECD 199, 200
ontology 172
ozone layer 122, 129, 130

paleoecology 47, 171
paradigm 3, 5, 7, 34, 38, 41, 189,
parks 33–54, 105–19
  national 36, 37, 43, 44, 66, 161, 164
parochialism 41, 134
participatory scenario investigation 35
pasture land management 181–95
pedagogy 186, 198–204
photography 140
pictorial perspective 138, 147
Picturesque 63, 69, 74, 84, 140, 142, 143
pluralism 8
poetry 60, 61, 64, 65, 71, 72
pollution 29, 99, 109, 133, 153
  air 10, 31, 95
  marine 155
  water 10, 31
positivism 34, 35
post-Fordism 19, 156
postmodernism 8, 171
post-normal science 34, 35, 41, 48, 91
postructuralism 171
powerful learning experiences 204
power of worldview 171–3

privatism 96, 97
problem orientation 34, 35
prose 60, 71, 79, 82, 83
public opinion 12, 123
public opinion polls 122
public participation 90–92, 96, 98, 102, 105–20, 211

radio 1, 121–34
reader's construction 75, 82–4
readership filter 82
reading image and text 168
recession 19, 24, 29
recreation 24, 105, 109–19, 207
reductionism 34, 41
relativism 41
resource cycle dynamics 19
resource management 33–53
resource peripheries 18, 19–21
rights
  aboriginal 20
  autonomist 156
  biotic 9
  group 153
  human 9
  international human 171
  hunting 58
  language-related 171
  native 153
  property 175
  treaty 24
Romanticism 63, 74

scientific rationalism 91, 92
self-efficacy 203
social process 35, 40, 445, 48
social science 5, 10, 18, 62, 172
spaceship earth 24
standpoint 35, 39, 45, 47
stereotypes 10, 145
stewardship 17, 105, 110, 116, 117, 119, 187, 190
sub-atomic physics 8

technocentrism 8, 9, 206
technology transfer 183, 186, 187
television 1, 3, 43, 65, 121–34, 208
text 1, 12, 69, 75, 82, 168
textual study 73
theory of operant conditioning 193
theory of planned behaviour 192, 193
theories-in-use 45
tourism 24, 47, 97, 140, 164
travel writing 71

UNESCO 199, 205, 210
union movement 28
utopianism 206

value judgements 4, 12, 34, 91, 101
values
  aesthetic 17, 23
  Arctic 167–80
  bourgeois 138, 142, 144
  common 9
  cross-disciplinary 5, 9
  cultural 8, 17, 33
  ecological 17, 23, 155
  emotive 17
  industrial 66
  imperialist 138
  Inuit 169
  moral 17
  non-industrial 19
  political 90
  rural 21
  urban 21
vocational training 200
volunteer groups 113, 114, 117, 118

wars in the woods 18, 28, 30–31
waste
  hazardous 33
  household 1
  nuclear 133
way-finding 69–87
writing construction 75, 79–82

The GeoJournal Library

1. B. Currey and G. Hugo (eds.): *Famine as Geographical Phenomenon.* 1984 ISBN 90-277-1762-1
2. S.H.U. Bowie, F.R.S. and I. Thornton (eds.): *Environmental Geochemistry and Health.* Report of the Royal Society's British National Committee for Problems of the Environment. 1985 ISBN 90-277-1879-2
3. L.A. Kosiński and K.M. Elahi (eds.): *Population Redistribution and Development in South Asia.* 1985 ISBN 90-277-1938-1
4. Y. Gradus (ed.): *Desert Development.* Man and Technology in Sparselands. 1985 ISBN 90-277-2043-6
5. F.J. Calzonetti and B.D. Solomon (eds.): *Geographical Dimensions of Energy.* 1985 ISBN 90-277-2061-4
6. J. Lundqvist, U. Lohm and M. Falkenmark (eds.): *Strategies for River Basin Management.* Environmental Integration of Land and Water in River Basin. 1985 ISBN 90-277-2111-4
7. A. Rogers and F.J. Willekens (eds.): *Migration and Settlement.* A Multiregional Comparative Study. 1986 ISBN 90-277-2119-X
8. R. Laulajainen: *Spatial Strategies in Retailing.* 1987 ISBN 90-277-2595-0
9. T.H. Lee, H.R. Linden, D.A. Dreyfus and T. Vasko (eds.): *The Methane Age.* 1988 ISBN 90-277-2745-7
10. H.J. Walker (ed.): *Artificial Structures and Shorelines.* 1988 ISBN 90-277-2746-5
11. A. Kellerman: *Time, Space, and Society.* Geographical Societal Perspectives. 1989 ISBN 0-7923-0123-4
12. P. Fabbri (ed.): *Recreational Uses of Coastal Areas.* A Research Project of the Commission on the Coastal Environment, International Geographical Union. 1990 ISBN 0-7923-0279-6
13. L.M. Brush, M.G. Wolman and Huang Bing-Wei (eds.): *Taming the Yellow River: Silt and Floods.* Proceedings of a Bilateral Seminar on Problems in the Lower Reaches of the Yellow River, China. 1989 ISBN 0-7923-0416-0
14. J. Stillwell and H.J. Scholten (eds.): *Contemporary Research in Population Geography.* A Comparison of the United Kingdom and the Netherlands. 1990 ISBN 0-7923-0431-4
15. M.S. Kenzer (ed.): *Applied Geography.* Issues, Questions, and Concerns. 1989 ISBN 0-7923-0438-1
16. D. Nir: *Region as a Socio-environmental System.* An Introduction to a Systemic Regional Geography. 1990 ISBN 0-7923-0516-7
17. H.J. Scholten and J.C.H. Stillwell (eds.): *Geographical Information Systems for Urban and Regional Planning.* 1990 ISBN 0-7923-0793-3
18. F.M. Brouwer, A.J. Thomas and M.J. Chadwick (eds.): *Land Use Changes in Europe.* Processes of Change, Environmental Transformations and Future Patterns. 1991 ISBN 0-7923-1099-3

The GeoJournal Library

---

19. C.J. Campbell: *The Golden Century of Oil 1950–2050.* The Depletion of a Resource. 1991 ISBN 0-7923-1442-5
20. F.M. Dieleman and S. Musterd (eds.): *The Randstad: A Research and Policy Laboratory.* 1992 ISBN 0-7923-1649-5
21. V.I. Ilyichev and V.V. Anikiev (eds.): *Oceanic and Anthropogenic Controls of Life in the Pacific Ocean.* 1992 ISBN 0-7923-1854-4
22. A.K. Dutt and F.J. Costa (eds.): *Perspectives on Planning and Urban Development in Belgium.* 1992 ISBN 0-7923-1885-4
23. J. Portugali: *Implicate Relations.* Society and Space in the Israeli-Palestinian Conflict. 1993 ISBN 0-7923-1886-2
24. M.J.C. de Lepper, H.J. Scholten and R.M. Stern (eds.): *The Added Value of Geographical Information Systems in Public and Environmental Health.* 1995 ISBN 0-7923-1887-0
25. J.P. Dorian, P.A. Minakir and V.T. Borisovich (eds.): *CIS Energy and Minerals Development.* Prospects, Problems and Opportunities for International Cooperation. 1993 ISBN 0-7923-2323-8
26. P.P. Wong (ed.): *Tourism vs Environment: The Case for Coastal Areas.* 1993 ISBN 0-7923-2404-8
27. G.B. Benko and U. Strohmayer (eds.): *Geography, History and Social Sciences.* 1995 ISBN 0-7923-2543-5
28. A. Faludi and A. der Valk: *Rule and Order. Dutch Planning Doctrine in the Twentieth Century.* 1994 ISBN 0-7923-2619-9
29. B.C. Hewitson and R.G. Crane (eds.): *Neural Nets: Applications in Geography.* 1994 ISBN 0-7923-2746-2
30. A.K. Dutt, F.J. Costa, S. Aggarwal and A.G. Noble (eds.): *The Asian City: Processes of Development, Characteristics and Planning.* 1994 ISBN 0-7923-3135-4
31. R. Laulajainen and H.A. Stafford: *Corporate Geography.* Business Location Principles and Cases. 1995 ISBN 0-7923-3326-8
32. J. Portugali (ed.): *The Construction of Cognitive Maps.* 1996 ISBN 0-7923-3949-5
33. E. Biagini: *Northern Ireland and Beyond.* Social and Geographical Issues. 1996 ISBN 0-7923-4046-9
34. A.K. Dutt (ed.): *Southeast Asia: A Ten Nation Region.* 1996 ISBN 0-7923-4171-6
35. J. Settele, C. Margules, P. Poschlod and K. Henle (eds.): *Species Survival in Fragmented Landscapes.* 1996 ISBN 0-7923-4239-9
36. M. Yoshino, M. Domrös, A. Douguédroit, J. Paszynski and L.D. Nkemdirim (eds.): *Climates and Societies – A Climatological Perspective.* A Contribution on Global Change and Related Problems Prepared by the Commission on Climatology of the International Geographical Union. 1997 ISBN 0-7923-4324-7
37. D. Borri, A. Khakee and C. Lacirignola (eds.): *Evaluating Theory-Practice and Urban-Rural Interplay in Planning.* 1997 ISBN 0-7923-4326-3

The GeoJournal Library

38. J.A.A. Jones, C. Liu, M-K. Woo and H-T. Kung (eds.): *Regional Hydrological Response to Climate Change.* 1996 ISBN 0-7923-4329-8
39. R. Lloyd: *Spatial Cognition.* Geographic Environments. 1997 ISBN 0-7923-4375-1
40. I. Lyons Murphy: *The Danube: A River Basin in Transition.* 1997 ISBN 0-7923-4558-4
41. H.J. Bruins and H. Lithwick (eds.): *The Arid Frontier.* Interactive Management of Environment and Development. 1998 ISBN 0-7923-4227-5
42. G. Lipshitz: *Country on the Move: Migration to and within Israel, 1948–1995.* 1998 ISBN 0-7923-4850-8
43. S. Musterd, W. Ostendorf and M. Breebaart: *Multi-Ethnic Metropolis: Patterns and Policies.* 1998 ISBN 0-7923-4854-0
44. B.K. Maloney (ed.): *Human Activities and the Tropical Rainforest.* Past, Present and Possible Future. 1998 ISBN 0-7923-4858-3
45. H. van der Wusten (ed.): *The Urban University and its Identity.* Roots, Location, Roles. 1998 ISBN 0-7923-4870-2
46. J. Kalvoda and C.L. Rosenfeld (eds.): *Geomorphological Hazards in High Mountain Areas.* 1998 ISBN 0-7923-4961-X
47. N. Lichfield, A. Barbanente, D. Borri, A. Khakee and A. Prat (eds.): *Evaluation in Planning.* Facing the Challenge of Complexity. 1998 ISBN 0-7923-4870-2
48. A. Buttimer and L. Wallin (eds.): *Nature and Identity in Cross-Cultural Perspective.* 1999 ISBN 0-7923-5651-9
49. A. Vallega: *Fundamentals of Integrated Coastal Management.* 1999 ISBN 0-7923-5875-9
50. D. Rumley: *The Geopolitics of Australia's Regional Relations.* 1999 ISBN 0-7923-5916-X
51. H. Stevens: *The Institutional Position of Seaports.* An International Comparison. 1999 ISBN 0-7923-5979-8
52. H. Lithwick and Y. Gradus (eds.): *Developing Frontier Cities.* Global Perspectives – Regional Contexts. 2000 ISBN 0-7923-6061-3
53. H. Knippenberg and J. Markusse (eds.): *Nationalising and Denationalising European Border Regions, 1800–2000.* Views from Geography and History. 2000 ISBN 0-7923-6066-4
54. R. Gerber and G.K. Chuan (eds.): *Fieldwork in Geography: Reflections, Perspectives and Actions.* 2000 ISBN 0-7923-6329-9
55. M. Dobry (ed.): *Democratic and Capitalist Transitions in Eastern Europe.* Lessons for the Social Sciences. 2000 ISBN 0-7923-6331-0
56. Y. Murayama: *Japanese Urban System.* 2000 ISBN 0-7923-6600-X
57. D. Zheng, Q. Zhang and S. Wu (eds.): *Mountain Geoecology and Sustainable Development of the Tibetan Plateau.* 2000 ISBN 0-7923-6688-3

The GeoJournal Library

58. A.J. Conacher (ed.): *Land Degradation.* Papers selected from Contributions to the Sixth Meeting of the International Geographical Union's Commission on Land Degradation and Desertification, Perth, Western Australia, 20–28 September 1999. 2001 ISBN 0-7923-6770-7
59. S. Conti and P. Giaccaria: *Local Development and Competitiveness.* 2001 ISBN 0-7923-6829-0
60. P. Miao (ed.): *Public Places in Asia Pacific Cities.* Current Issues and Strategies. 2001 ISBN 0-7923-7083-X
61. N. Maiellaro (ed.): *Towards Sustainable Buiding.* 2001 ISBN 1-4020-0012-X
62. G.S. Dunbar (ed.): *Geography: Discipline, Profession and Subject since 1870.* An International Survey. 2001 ISBN 1-4020-0019-7
63. J. Stillwell and H.J. Scholten (eds.): *Land Use Simulation for Europe.* 2001 ISBN 1-4020-0213-0
64. P. Doyle and M.R. Bennett (eds.): *Fields of Battle.* Terrain in Military History. 2002 ISBN 1-4020-0433-8
65. C.M. Hall and A.M. Williams (eds.): *Tourism and Migration.* New Relationships between Production and Consumption. 2002 ISBN 1-4020-0454-0
66. I.R. Bowler, C.R. Bryant and C. Cocklin (eds.): *The Sustainability of Rural Systems.* Geographical Interpretations. 2002 ISBN 1-4020-0513-X
67. O. Yiftachel, J. Little, D. Hedgcock and I. Alexander (eds.): *The Power of Planning.* Spaces of Control and Transformation. 2001 ISBN Hb; 1-4020-0533-4 ISBN Pb; 1-4020-0534-2
68. K. Hewitt, M.-L. Byrne, M. English and G. Young (eds.): *Landscapes of Transition.* Landform Assemblages and Transformations in Cold Regions. 2002 ISBN 1-4020-0663-2
69. M. Romanos and C. Auffrey (eds.): *Managing Intermediate Size Cities.* Sustainable Development in a Growth Region of Thailand. 2002 ISBN 1-4020-0818-X
70. B. Boots, A. Okabe and R. Thomas (eds.): *Modelling Geographical Systems.* Statistical and Computational Applications. 2003 ISBN 1-4020-0821-X
71. R. Gerber and M. Williams (eds.): *Geography, Culture and Education.* 2002 ISBN 1-4020-0878-3
72. D. Felsenstein, E.W. Schamp and A. Shachar (eds.): *Emerging Nodes in the Global Economy: Frankfurt and Tel Aviv Compared.* 2002 ISBN 1-4020-0924-0
73. R. Gerber (ed.): *International Handbook on Geographical Education.* 2003 ISBN 1-4020-1019-2
74. M. de Jong, K. Lalenis and V. Mamadouh (eds.): *The Theory and Practice of Institutional Transplantation.* Experiences with the Transfer of Policy Institutions. 2002 ISBN 1-4020-1049-4
75. A.K. Dutt, A.G. Noble, G. Venugopal and S. Subbiah (eds.): *Challenges to Asian Urbanization in the 21st Century.* 2003 ISBN 1-4020-1576-3

The GeoJournal Library

76. I. Baud, J. Post and C. Furedy (eds.): *Solid Waste Management and Recycling.* Actors, Partnerships and Policies in Hyderabad, India and Nairobi, Kenya. 2004 ISBN 1-4020-1975-0
77. A. Bailly and L.J. Gibson (eds.): *Applied Geography.* A World Perspective. 2004 ISBN 1-4020-2441-X
78. H.D. Smith (ed.): *The Oceans: Key Issues in Marine Affairs.* 2004 ISBN 1-4020-2746-X
79. M. Ramutsindela: *Parks and People in Postcolonial Societies.* Experiences in Southern Africa. 2004 ISBN 1-4020-2542-4
80. R.A. Boschma and R.C. Kloosterman (eds.): *Learning from Clusters.* A Critical Assessment from an Economic-Geographical Perspective. 2005 ISBN 1-4020-3671-X
81. G. Humphrys and M. Williams (eds.): *Presenting and Representing Environments.* 2005 ISBN 1-4020-3813-5